Success guides

Standard Grade
Maths

Nicola Izatt ✗ Fiona Mapp

Contents

Number and Money

Types of number	4
Positive and negative numbers	6
Fractions	8
Decimals	10
Percentages 1	12
Percentages 2	14
Equivalence	16
Using a calculator	17
Calculations	18
Ratios	20
Proportion	22
Money	24
Speed, distance, time	26
Standard index form	28
Test your progress	30

Algebra

Algebra 1	32
Algebra 2	34
Number patterns and sequences	35
Equations	36
Inequalities	38
Straight line graphs	40
Interpreting graphs	42
Test your progress	44

Measure and Shape

Metric units	46
2D shapes	47
3D	48
Nets	50
Symmetry	51
Circles	52

Measure and Shape (continued)

Angles	54
Bearings and scale drawings	56
Similarity	58
Sample exam-style questions	59
Pythagoras theorem	60
Trigonometry in right-angled triangles	62
Applications of trigonometry	64
Test your progress	66

Graphs and Statistics

Representing data	68
Scatter diagrams and correlation	70
Averages	72
Probability 1	74
Probability 2	76
Test your progress	78

Non-Calculator Skills

Whole numbers	80
Test your progress	82

Example questions

Number and money	84
Algebra	86
Measure and shape	88
Graphs and statistics	90

Mixed questions	92
Answers	94
Index	96

Types of numbers

Squares and cubes

Square numbers

Anything to the **power 2** is **square**.

For example, $6^2 = 6 \times 6 = 36$ (six squared).

Square numbers include:

   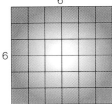

1	4	9	16	25	36	49	64	81	100	...
(1×1)	(2×2)	(3×3)	(4×4)	(5×5)	(6×6)	(7×7)	(8×8)	(9×9)	(10×10)	

Square numbers can be illustrated by drawing squares:

  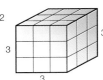 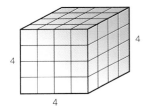

Cube numbers

Anything to the **power 3** is **cube**.

For example, $5^3 = 5 \times 5 \times 5 = 125$ (five cubed).

Cube numbers include:

1	8	27	64	125	216 ...
$(1 \times 1 \times 1)$	$(2 \times 2 \times 2)$	$(3 \times 3 \times 3)$	$(4 \times 4 \times 4)$	$(5 \times 5 \times 5)$	$(6 \times 6 \times 6)$

Cube numbers can be illustrated by drawing cubes:

Top Tip

Remember when you square any number your answer is always positive. This is important to remember when doing non–calculator work. e.g. $(-3)^2 = 9$.

Square roots and cube roots

$\sqrt{}$ is the square root sign. Taking the square root is the opposite of squaring.

For example, $\sqrt{25} = \pm 5$ since $5^2 = 25$, or $(-5)^2 = 25$.

$\sqrt[3]{}$ is the cube root sign. Taking the cube root is the opposite of cubing.

For example, $\sqrt[3]{8} = 2$ since $2^3 = 8$.

Multiples

These are just the numbers in multiplication tables.

For example, multiples of 8 are 8, 16, 24, 32, 40, ...

Factors

These are numbers which divide exactly into another number.

For example, the factors of 20 are 1, 2, 4, 5, 10, 20.

Factors of 20 can be split up into factor pairs.

E.g: 1×20, 2×10, 4×5

Prime numbers

These are numbers which have only two factors, 1 and themselves.

Note that 1 is not a prime number and 2 is the only even prime.

Prime factors

These are factors which are prime.

All numbers, except prime numbers, can be written as products of their prime factors.

Example

The diagram on the right shows the prime factors of 360.

- Divide 360 by its first prime factor, 2.
- Divide 180 by its first prime factor, 2.
- Keep on going until the final number is prime.

As a **product of its prime factors** 360 may be written as:

$2 \times 2 \times 2 \times 3 \times 3 \times 5 = 360$

Highest common factor (HCF)

The **largest factor** that two numbers have in common is called the **HCF**.

Example

Find the HCF of 84 and 360.

- Write the numbers as products of their prime factors.

$$84 = 2 \times 2 \times \quad 3 \quad \times 7$$
$$360 = 2 \times 2 \times 2 \times 3 \times 3 \times 5$$

- Ring the factors in common.
- These give the HCF $= 2 \times 2 \times 3 = 12$.

Lowest common multiple (LCM)

This is the **lowest** number which is a **multiple** of two numbers.

Example

Find the LCM of 6 and 8.

- Write the numbers as products of their prime factors.
- $8 = 2 \times 2 \times 2$
- $6 = \quad 2 \times 3$
- 8 and 6 have a common prime factor of 2. So it is only counted once.
- The LCM of 6 and 8 is $2 \times 2 \times 2 \times 3 = 24$.

Quick Test

1. Find the HCF and LCM of 24 and 60.
2. Find **a)** 12^2 **b)** $\sqrt{81}$
3. List the prime factors of 90.

Answers: 1. HCF – 12   LCM – 120. 2. a) 144 b) 9 3. $2 \times 3 \times 3 \times 5$

Positive & negative numbers

Directed numbers

These are numbers which may be **positive** or **negative**.
Positive numbers are above zero, negative numbers are below zero.

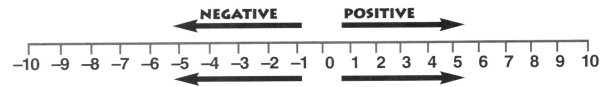

NEGATIVE   **POSITIVE**

–10 –9 –8 –7 –6 –5 –4 –3 –2 –1 0 1 2 3 4 5 6 7 8 9 10

GETTING SMALLER   **GETTING BIGGER**

Examples

–10 is smaller than –8   –4 is bigger than –8   2 is bigger than –6

–10 < –8   **–4 > –8**   **2 > –6**

Directed numbers are often seen on the weather forecast in winter.

Quite often the temperature is below 0.

Aberdeen is the coldest at –8°C.

London is 6°C warmer than Manchester.

Integers

The integers are the set of whole numbers including zero {..., –3, –2, –1, 0, 1, 2, 3, ...}.

When referring to integers, the term **integral value** is used. A number that is **non–integral** is not an integer.

Multiplying and dividing directed numbers

Multiply and divide the numbers as normal.
Then find the sign for the answer using these rules:

- two **like** signs (both + or both –) give positive,
- two **unlike** signs (one + and the other –) give negative.

Examples

–6 × (+4) = –24

–12 ÷ (–3) =   4

–6 × (–3) =   18

20 ÷ (–4) =   –5

Top Tip
The rules of multiplication and division need to be memorised.

(+) × (+) = +
(–) × (–) = +
(+) × (–) = –
(–) × (+) = –

(+) ÷ (+) = +
(–) ÷ (–) = +
(+) ÷ (–) = –
(–) ÷ (+) = –

Adding and subtracting directed numbers

Example

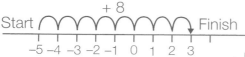

Start ... + 8 ... Finish

−5 −4 −3 −2 −1 0 1 2 3

The temperature at 6 a.m. was −5°C. By 10 a.m. it had risen 8 degrees. So the new temperature was 3°C.

Example

Find the value of −2 − 4

−2 − 4.

This represents the **sign** of the number. Start at −2.

This represents the operation of **subtraction**. Move 4 places to the left.

Finish ... −4 ... Start

−6 −5 −4 −3 −2 −1 0 1 2 3 4

So −2 − 4 = −6

Top Tip
You can always draw a number line to help with the addition & subtraction of negative numbers.

When the number to be added (or subtracted) is **negative**, the normal direction of movement is **reversed**.

Example

−4 − (−3) is the same as
−4 + 3 = −1

The negative changes the **direction**. Move 3 places to the **right**.

When two (+) or two (−) signs are together, these rules are used:

+ (+) → + ⎫ **like** signs give
− (−) → + ⎭ a **positive**

− (+) → − ⎫ **unlike** signs give
+ (−) → − ⎭ a **negative**

Examples

−6 + (−2) = −6 − 2 = −8
−2 − (+6) = −2 − 6 = −8
4 − (−3) = 4 + 3 = 7
9 + (−3) = 9 − 3 = 6

Negative numbers on the calculator

The +/− or (−) key on the calculator gives a **negative** number.
This represents the **sign**.

For example, to get −6, press
 or .

Example

−4 − (−2) = −2 is keyed in the calculator like this:

sign   operation   sign

Make sure you know how to enter negative numbers in your own calculator.

Quick Test

1. If the temperature was −12°C at 2 a.m., and it rose by 15 degrees by 11 a.m., what was the temperature at 11 a.m.?

2. Work these out, without a calculator.
   a) −2 − (−6)   b) −9 + (−7)   c) −2 × 6   d) −9 + (−3)   e) −20 ÷ (−4)
   f) −18 ÷ (−3)   g) 4 − (−3)   h) −7 + (−3)   i) −9 × −4

Fractions

A fraction is a part of a whole one. $\frac{4}{5}$ means 4 parts out of 5.
The top number is the **numerator**. The bottom number is the **denominator**.
A fraction like $\frac{4}{5}$ is called a **proper fraction**.
A fraction like $\frac{24}{17}$ is called an **improper fraction**.
$2\frac{1}{2}$ is called a **mixed number**.

Equivalent fractions

These are fractions which have the same value.

Example

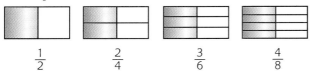

$\frac{1}{2}$ $\frac{2}{4}$ $\frac{3}{6}$ $\frac{4}{8}$

From the diagrams it can be seen that $\frac{1}{2} = \frac{2}{4} = \frac{3}{6} = \frac{4}{8}$.
They are equivalent fractions. Fractions can be changed to their equivalent by multiplying or dividing both the numerator and denominator by the same amount.

Examples

$\frac{5}{7} \overset{\times 4}{\underset{\times 4}{\rightleftarrows}} \frac{20}{28}$

Change $\frac{5}{7}$ to its equivalent fraction with a denominator of 28.
Multiply top and bottom by 4.
So $\frac{20}{28}$ is equivalent to $\frac{5}{7}$.

 $\frac{40}{60} \overset{\div 20}{\underset{\div 20}{\rightleftarrows}} \frac{2}{3}$

Change $\frac{40}{60}$ to its equivalent fraction with a denominator of 3.
Divide top and bottom by 20.
So $\frac{40}{60}$ is equivalent to $\frac{2}{3}$.

Using the fraction key on the calculator

$\boxed{a^b/_c}$ is the fraction key on the calculator.

Example
$\frac{12}{18}$ is keyed in as

$\boxed{1}$ $\boxed{2}$ $\boxed{a^b/_c}$ $\boxed{1}$ $\boxed{8}$

This is displayed as $\boxed{12 \lrcorner 18}$
or $\boxed{12 \ulcorner 18}$

The calculator will automatically reduce down fractions when the $\boxed{=}$ key is pressed.
For example, $\frac{12}{18}$ becomes

$\boxed{2 \lrcorner 3}$ or $\boxed{2 \ulcorner 3}$ **This means two-thirds.**

A display of $\boxed{1 \lrcorner 4 \lrcorner 9}$
means $1\frac{4}{9}$.
If you now press $\boxed{\text{shift}}$ $\boxed{a^b/_c}$, it converts back to an improper fraction, $\boxed{13 \ulcorner 9}$

Multiplication and division of fractions

When multiplying and dividing fractions, write out whole or mixed numbers as improper fractions before starting.

Example   — Multiply numerators together.
$\frac{2}{9} \times \frac{4}{7} = \frac{2 \times 4}{4 \times 7} = \frac{8}{63}$ — Multiply denominators together.

Change a **division** into a multiplication by turning the second fraction upside down and multiplying both fractions together; that is, **multiply by the reciprocal**.

Example
$\frac{7}{9} \div \frac{12}{18} = \frac{7}{9} \times \frac{18}{12} = \frac{126}{108} = 1\frac{1}{6}$ — Rewrite the answer as a mixed number.

Take the reciprocal of the second fraction.

Addition and subtraction of fractions

These examples show the basic principles of adding and subtracting fractions.

Example

$\frac{1}{8} + \frac{3}{4}$ • First make the denominators the same: $\frac{3}{4} = \frac{6}{8}$

$\frac{3}{4}$ is equivalent to $\frac{6}{8}$.

$= \frac{1}{8} + \frac{6}{8}$ • Replace $\frac{3}{4}$ with $\frac{6}{8}$ so that the denominators are the same.

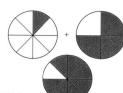

$= \frac{7}{8}$ • Add the numerators $1 + 6 = 7$.
Do not add the denominators; the denominator stays the same number.

Example

$\frac{9}{12} - \frac{1}{3}$ • First make the denominators the same: $\frac{1}{3} = \frac{4}{12}$

$\frac{1}{3}$ is equivalent to $\frac{4}{12}$.

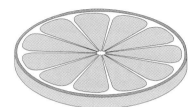

$= \frac{9}{12} - \frac{4}{12}$ • Replace the $\frac{1}{3}$ with $\frac{4}{12}$.

$= \frac{5}{12}$ • Subtract the numerators but not the denominators; the denominator stays the same number.

On a calculator you would key in:

Fraction of a quantity

This example shows the basic principles of finding a fraction of a quantity.

Example:

$\frac{4}{5}$ of 30m

• First find $\frac{1}{5}$ (divide 30m by 5): 6m

• $\frac{4}{5}$ is 4 times as much: 24m.

Top Tip
Questions involving fractions will always appear in the non-calculator paper.

Quick Test

1. Without using a calculator work out the following:

   a) $\frac{2}{9} + \frac{3}{27}$ b) $\frac{3}{5} - \frac{1}{4}$ c) $\frac{6}{9} \times \frac{72}{104}$ d) $\frac{8}{9} \div \frac{2}{3}$

   e) $\frac{4}{7} - \frac{1}{3}$ f) $\frac{2}{7} \div 1\frac{1}{2}$ g) $\frac{7}{11} \div \frac{22}{14}$ h) $\frac{2}{9} + \frac{4}{7}$

2. Calculate $\frac{2}{9}$ of £180.

3. Calculate $\frac{5}{6}$ of an hour.

Decimals

A decimal point is used to separate whole number columns from fractional columns.

Example

Thousands	Hundreds	Tens	Units	.	Tenths	Hundredths	Thousandths
5	9	2	4	.	1	6	3

Decimal Point

The 1 means $\frac{1}{10}$. The 6 means $\frac{6}{100}$. The 3 means $\frac{3}{1000}$.

Recurring decimals

A decimal that recurs is shown by placing a dot over the numbers that repeat.

Examples

0.333... = 0.$\dot{3}$ 0.17777... = 0.1$\dot{7}$ 0.232323... = 0.$\dot{2}\dot{3}$

Decimal places (d.p.)

When rounding numbers to a specified number of decimal places (d.p.):

- look at the last number that is wanted (if rounding 12.367 to 2 d.p., look at the 6 which is the second d.p.)
- look at the number to the right of it (the 7)
- if it is 5 or more, then round up the last digit (7 is greater than 5, so round up the 6 to a 7)
- if it is less than 5, then the digit remains the same.

Examples

Round 12.49 to 1 d.p.
12.4<u>9</u> rounds up to 12.5.

Round 8.735 to 2 d.p.
8.73<u>5</u> rounds up to 8.74.

Round 9.624 to 2 d.p.
9.62<u>4</u> rounds to 9.62

Ordering decimals

When ordering decimals:

- first write them with the same number of figures after the decimal point by adding zeros;
- then compare whole numbers, digits in the tenths place, digits in the hundredths place, and so on.

Examples

Arrange these numbers in order of size, smallest first:

6.21, 6.023, 6.4, 6.04, 2.71, 9.4

First rewrite them:

6.210, 6.023, 6.400, 6.040, 2.710, 9.400

Then re-order them:

2.710, 6.023, 6.040, 6.210, 6.400, 9.400

Remember, hundredths are smaller than tenths $\frac{10}{100} = \frac{1}{10}$ so $\frac{6}{100} < \frac{1}{10}$

Calculations with decimals

When adding and subtracting decimals, the decimal points need to go under each other.

Examples

Line up the digits carefully.

$$27.46$$
$$7.291 +$$
$$\overline{34.751}$$
$$11$$

Put the decimal points under each other.

$$1\overset{6}{\cancel{7}}\overset{9}{.}\overset{1}{\cancel{0}}0$$
$$12.84 -$$
$$\overline{4.16}$$

The decimal point in the answer will be in line.

When **multiplying** decimals, the answer must have the same number of decimal places as the **total number of decimal places** in the numbers which are being multiplied.

Examples

$$24.6$$
$$7 \times$$
$$\overline{172.2}$$
$$3\;4$$

Multiply 246 by 7, ignoring the decimal point. 24.6 has 1 number after the decimal point. The answer must have 1 decimal place (1 d.p.).

So 24.6 x 7 = 172.2

Remember to check with your calculator.

Work out 4.52 x 0.2

$$452$$
$$2 \times$$
$$\overline{904}$$
$$1$$

Work out 452 x 2, ignoring the decimal points. 4.52 has 2 d.p. and 0.2 has 1 d.p. So the answer must have 3 d.p.

904 → 0.904 Move the decimal point 3 places.

So 4.52 x 0.2 = 0.904

Example

$$\begin{array}{r} 4.8 \\ 3\,\overline{\smash{)}1\overset{1}{4}.\overset{2}{4}} \end{array}$$

Put the decimal points in line.

When dividing decimals, divide as normal, placing the decimal points in line.

Multiplying and dividing by numbers between 0 and 1

When **multiplying** by numbers between 0 and 1, the **result is smaller** than the starting value.
When **dividing** by numbers between 0 and 1, the **result is bigger** than the starting value.

Examples

$6 \times 0.1 = 0.6$
$6 \times 0.01 = 0.06$
$6 \times 0.001 = 0.006$
The result is smaller than the starting value.

$6 \div 0.1 = 60$
$6 \div 0.01 = 600$
$6 \div 0.001 = 6000$
The result is bigger than the starting value.

Quick Test

1. Without using a calculator work out the following:
   a) 27.16 + 9.32   b) 29.04 − 11.361   c) 12.8 × 2.1   d) 49.2 ÷ 4
   e) 600 × 0.01   f) 520 × 0.1   g) 20 × 0.02   h) 37 × 0.0001
   i) 400 ÷ 0.1   j) 450 ÷ 0.01   k) 470 ÷ 0.001   l) 650 ÷ 0.02

2. Round the following numbers to 2 decimal places:
   a) 7.469   b) 12.0372   c) 9.365   d) 10.042   e) 8.1794

Percentages 1

Percentages are fractions with a denominator of 100

% is the percentage sign.

75% means $\frac{75}{100}$ (this is also equal to $\frac{3}{4}$).

75%

Percentage of a quantity

The word 'of' means **multiply**. For example, 40% of £600 becomes $\frac{40}{100} \times 600 = £240$

On the calculator, key in 4 0 ÷ 1 0 0 × 6 0 0 =

If this is on the non-calculator paper

- Work out 10% first by dividing by 10 (10% = $\frac{10}{100} = \frac{1}{10}$)
  e.g. 600 ÷ 10 = £60.

- Multiply by 4 to get 40% e.g. 4 × 60 = £240.

Example

A meal for four costs £92.20. VAT is charged at 17.5%.
(VAT is a tax which is added on to the cost of most items.)

a) How much VAT is there to pay on the meal?

b) What is the final price of the meal?

a) 17.5% of £92.20 = $\frac{17.5}{100} \times 92.20$ = £16.14 (to the nearest penny)
VAT = £16.14

b) Price of meal = £92.20 + £16.14 = £108.34

An alternative is to use a scale factor method.

- an increase of 17.5%, is the same as multiplying by $1 + \frac{17.5}{100} = 1.175$.

£92.20 x 1.175 = £108.34 (to the nearest penny).

Top Tip
Percentage questions appear frequently on Standard Grade non-calculator papers. It sometimes helps to work out what 10% is equal to, as shown in the example above.

This is just like a percentage of a quantity question.

Percentage increase and decrease

If the answer required is a percentage, then you will need to multiply by 100%.

$$\% \text{ change} = \frac{\text{change}}{\text{original}} \times 100\%$$

Example

A coat costs £125. In a sale it is reduced to £85.

What is the percentage reduction?

Reduction = £125 – £85 = £40

$\frac{40}{125} \times 100\% = 32\%$

£125 £85

Example

Matthew bought a flat for £45000.
Three years later, he sold it for £62000.
What was his percentage profit?

Profit = £62000 – £45000
        = £17000

% Profit = $\frac{17000}{45000} \times 100\%$
          = 37.78%

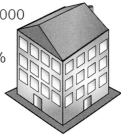

One quantity as a percentage of another

To make the answer a **percentage**, multiply by **100%**.

Example

In a carton of milk, 6.2 g of the contents are fat.
If 2.5 g of the fat is saturated, what percentage is this?
$\frac{2.5}{6.2} \times 100\% = 40.3\%$ (to 1 d.p.)

On the calculator key in

`2` `.` `5` `÷` `6` `.` `2` `×` `1` `0` `0` `=`

Fraction $\xrightarrow{\times 100\%}$ Percentage

Multiply the fraction by 100% to get a percentage.

Percentage problems: increase and decrease

Many exam questions ask you to find the increased or decreased value of a quantity.

Example

The price of a television is reduced by 30% in a summer sale.
Before the sale the television cost £450.

a. What is the discount?

30% of £450 = $\frac{30}{100} \times$ £450 = £135

b. What is the sale price of the television?

£450 − £135 = £315

Example

A bar of chocolate increases in size by 25%. There were 16 pieces in the original bar.

a. How many extra pieces are there?

25% of 16 = 0.25 × 16 = 4 pieces

b. How many pieces are in the bigger bar?

16 + 4 = 20 pieces

Quick Test

1. Work out 30% of £700.

2. Sarah got 94 out of 126 in a Maths test. What percentage did she get?

3. Reece weighed 6lb when he was born. If his weight has increased by 65%, how much does he now weigh?

4. Phil bought a car for £5600. Two years later he sold it for £4480. What is the percentage reduction in price?

5.
Super's football boots	Joe's football boots
$\frac{1}{3}$ off	28% off

If a pair of football boots costs £49.99, which shop sells them cheaper in the sale and what is the price?

Answers 1. £210 **2.** 74.6% **3.** 9.9lb **4.** 20% **5.** £33.33: Super's

Percentages 2

There are really only 2 types of percentage questions:

1. 'Percentage of'
Here you are given the percentage so you ÷ 100

2. Writing your answer as a percentage
Here you need to work out a percentage so you × 100

Simple interest

This is the interest that is sometimes paid on money in banks and building societies.
The interest is paid each year (**per annum** or **p.a.**) and is the same amount each year.

Example

Jonathan has £2500 in his savings account.
Simple interest is paid at 4.4% p.a.

How much does he have in his account at the end of the year?

100% + 4.4% = 104.4%
(increasing by 4.4 % is the same as multiplying by 100 + .044 = 104.4 %)

Total savings = $\frac{104.4}{100}$ × £2500 = £2610
Interest paid = £2610 − £2500 = £110

Example

Chris has £1200 in his savings account. Simple interest is paid at 3.5% p.a. (per annum).
How much interest does he get after 5 months?

This is the interest after a year:
$\frac{3.5}{100}$ × £1200 = £42

To work out the interest for 5 months, calculate the monthly interest then multiply by 5.

One month's interest:
£42 ÷ 12 = £3.50
Interest after 5 months:
£3.50 × 5 = £17.50.

Sample exam-style questions

1. Marie is joining her local gym. She sees the sign shown here. Calculate the cost for a year.

Answer

36 × 12 = £432
20% of £432 = 0.2 × 432 = £86.40

Discounted membership
£432 − £86.40 = £345.60

Membership

£36 per month

20% Discount when you buy membership for one year.

2. One mini disc costs £4.00.
On special offer is a set of 6 costing £18.

a. How much is saved by buying the set?

Answer

Cost for buying 6 individual mini discs   6 × £4.00 = £24
Amount saved   £24 − £18 = £6

b. Express the saving as a percentage cost of 6 single mini discs.

Answer

$\frac{6}{24}$ × 100%
The saving was 25%

Top Tip
You will have seen from examples and problems that there are a number of ways of performing percentage calculations. There is not a right or a wrong method – use the method that you feel most comfortable with.

Tax and National Insurance

National Insurance

National Insurance (NI) is usually deducted from a wage as a percentage of the wage.

Example

Sue earns £1402.65 a month. National Insurance at 9% is deducted. How much NI must she pay?

9% of £1402.65 = 0.09 x £1402.65 = £126.24

Income Tax

A percentage of a wage or salary is removed as income tax.

Personal allowances must first be removed in order to obtain the taxable income.

Example

Harold earns £190 per week. His first £62 is not taxable but the remainder is taxed at 24%. How much income tax does he pay each week?

Taxable income = £190 − £62 = £128
24% tax = 0.24 × £128 = £30.72
Tax per week = £30.72

Top Tip
Being able to answer questions like the examples shown in this section is important – not only because they appear on the examination paper but because you will come across them in everyday life. Most of the examples are 'percentage of' questions.

Quick Test

1. Charlotte has £4250 in the bank. If the interest rate is 6.8% p.a., how much interest on her savings will she get at the end of the year?

2. Louise earns £2100 a month. National Insurance at 12% is deducted from her wage. How much NI must she pay?

3. A flat was bought in 1998 for £62000. In 1999 the price increased by 20% and then by a further 35% in 2000. How much was the flat worth at the end of 2000?

4. Fiona has £3200 savings. The simple interest is 3%. How much interest will she have after 4 months?

Answers 1. £289 2. £252 3. £100440 4. £32

Equivalences between fractions, decimals and percentages

Fractions to decimals to percentages

Fractions, decimals and percentages all mean the same thing but are just written in a different way.

Fraction	Decimal	Percentage
$\frac{1}{2}$	0.5	50%
$\frac{1}{3}$	0.3̇	33.3̇%
$\frac{2}{3}$	0.6̇	66.6̇%
$\frac{1}{4}$	0.25	25%
$\frac{3}{4}$	0.75	75%
$\frac{1}{5}$	0.2	20%
$\frac{1}{8}$	0.125	12.5%
$\frac{3}{8}$	0.375	37.5%
$\frac{1}{10}$	0.1	10%
$\frac{1}{100}$	0.01	1%

$3 \div 4 \longrightarrow$ (to 0.75) $\times 100\% \longrightarrow$ (to 75%)

The above table shows

- Some common fractions and their equivalents which you need to learn.
- How to convert fractions $\longrightarrow$ decimals $\longrightarrow$ percentages.

Ordering different numbers

When putting fractions, decimals and percentages in order of size, it is best to change them all to decimals first.

Example Place in order of size, smallest first, the following:

$\frac{3}{5}$, 0.65, 0.273, 27%, 62%, $\frac{4}{9}$

0.6, 0.65, 0.273, 0.27, 0.62, 0.4̇4̇ $\longleftarrow$ Convert fractions into decimals first

0.27, 0.273, 0.4̇4̇, 0.6, 0.62, 0.65 $\longleftarrow$ Now order

Top Tip
Get a friend to test you on the equivalences between fractions, decimals and percentages because you need to learn them.

Quick Test

1. Change the following fractions into
   a) decimals b) percentages
   i) $\frac{2}{7}$ ii) $\frac{3}{5}$ iii) $\frac{8}{9}$

2. Place in order of size, smallest first:
   $\frac{2}{5}$, 0.42, 0.041, $\frac{1}{3}$, 5%, 26%

Using a calculator

Important calculator keys

This picture shows you some of the most important keys on a calculator.
Make sure you are familiar with your own calculator.

> Practise using your own calculator. Make sure you know where these keys are.

Shift or 2nd or Inv allow 2nd functions to be carried out.

allows a fraction to be put in the calculator.

– or +/– changes positive numbers to negative ones.

bracket keys

often puts the ×10 part in when working in standard form.

square root

square button

trigonometric buttons

memory keys

works out powers

cancels only the last key you have pressed

all memory keys

[Calculator keys shown: SHIFT, √, x^2, $a\ ^b/_c$, SIN, COS, TAN, +−, MIN, MR, [(,)], x^y, 7, 8, 9, AC, 4, 5, 6, C, 1, 2, 3, ., 0, EXP (π), =, M+]

pressing **SHIFT** **EXP** often gives π

Calculating powers

 or is used for calculating powers such as 2^7.

- Use the power key on the calculator to work out 2^7.
- Write down calculator keys used.
- Check that you obtain the answer 128.

Now try writing down the keys that would be needed for these calculations.
Check that you get the right answers.

a) $\dfrac{2.9 \times 3.6}{(4.2 + 3.7)} = 1.322$

b) $9^{\frac{4}{9}} \times 4^5 = 2130$

c) $\dfrac{3 \times (5.2)^2}{9.6 \times (12.4)^3} = 4.432 \times 10^{-3}$

Example
$\dfrac{15 \times 10 + 46}{9.3 \times 2.1} = 10.04$ (2 d.p.)

This may be keyed in as:

[(... 15 × 10 + 46 ...)]
÷ [(... 9.3 × 2.1 ...)] =

The above could be as easily done using the memory keys. Try writing down the key sequence for yourself.

Quick Test

Work these out on your calculator:

a) $\dfrac{27.1 \times 6.4}{9.3 + 2.7}$

b) $\dfrac{(9.3)^4}{2.7 \times 3.6}$

c) $\sqrt{\dfrac{25^2}{4\pi}}$

d) $\dfrac{5}{9}(25 - 10)$

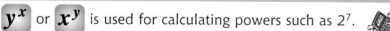

Top Tip
Make sure you know how to use the power key as it saves lots of time.

Answers a) 14.45 (2 d.p.) b) 769.6 (1 d.p.) c) 7052 (3 d.p.) d) $8\frac{1}{3}$ or 8.3

Approximating & checking calculations

Significant figures (s.f. or sig. fig.)

Apply the same rule as with decimal places: if the next digit is 5 or more, round up.

The 1st significant figure is the 1st digit which is not zero. The 2nd, 3rd, 4th, ... significant figures follow on after the 1st digit. They may or may not be zeros.

Take care when rounding that you do not change the place values.

Examples

6.4027 has 5 s.f.
1st 2nd 3rd 4th 5th

0.0004701 has 4 s.f.
1st 2nd 3rd 4th

Examples

Number	to 3 s.f.	to 2 s.f.	to 1 s.f.
4.207	4.21	4.2	4
4379	4380	4400	4000
0.006209	0.00621	0.0062	0.006

After rounding the last digit, you must fill in the end zeros.
For example,
4380 = 4400 to 2 s.f. (not 44).

Estimates and approximations

Estimating is a good way of checking answers.
- Round the numbers to 'easy' numbers, usually 1 or 2 significant figures.
- Work out the estimate using these easy numbers.
- Use the symbol $\approx$, which means '**approximately equal to**'.

For multiplying or dividing, never approximate a number with zero. Use 0.1, 0.01, 0.001, etc.

Examples

a) $8.93 \times 25.09 \approx 10 \times 25 = 250$
b) $(6.29)^2 \approx 6^2 = 36$
c) $\frac{296 \times 52.1}{9.72 \times 1.14} \approx \frac{300 \times 50}{10 \times 1} = \frac{15000}{10} = 1500$
d) $0.096 \times 79.2 \approx 0.1 \times 80 = 8$

Example

Jack does the calculation $\frac{9.6 \times 103}{(2.9)^2}$
a) Estimate the answer to this calculation, without using a calculator.
b) Jack's answer is 1175.7. Is this the right order of magnitude?
a) Estimate $\frac{9.6 \times 103}{(2.9)^2} \approx \frac{10 \times 100}{3^2} = \frac{1000}{9} \approx \frac{1000}{10} = 100$
b) Jack's answer is not the right order of magnitude. It is 10 times too big.

Right order of magnitude means 'about the right size'

When adding and subtracting, very small numbers may be approximated to zero.

Examples

$109.6 + 0.0002 \approx 110 + 0 = 110$ $63.87 - 0.01 \approx 64 - 0 = 64$

Top Tip
You will lose marks if you do not follow the exam instructions to round your answer to a certain number of significant figures or decimal places. Make sure you read the exam questions carefully to watch out for these instructions.

Calculations

When solving problems the answers should be rounded sensibly.

Example

$95.26 \times 6.39 = 608.7114 = 608.71$ (2 d.p.)

Example

Jackie has £9.37. She divides it equally between 5 people. How much does each person receive?

£9.37 ÷ 5 = £1.874
           = £1.87

Round to 2 d.p. because the values in the question are to 2 d.p.

Round to 1.87 because it is money.

Example

Paint is sold in 8 litre tins. Sandra needs 27 litres of paint. How many tins must she buy?

$27 \div 8 = 3$ remainder 3

Sandra needs 4 tins of paint.

Sandra would not have enough paint with 3 tins, since she is 3 litres short. Hence the number of tins of paint must be rounded up.

When rounding remainders, consider the context of the question.

Checking calculations

When checking calculations, the process can be reversed like this.

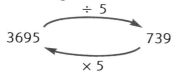

÷ 5

3695   739

× 5

Example

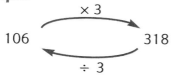

× 3

106   318

÷ 3

$106 \times 3 = 318$
Check: $318 \div 3 = 106$

Top Tip

You will lose marks if you do not write money to 2 d.p. If the answer to a money calculation is £9.7, **always** write it to 2 d.p. as £9.70.

Quick Test

1. Round the following numbers to 3 significant figures (3 s.f.):
   a) 0.003786
   b) 27490
   c) 307250

2. Estimate the answer to $\frac{(29.4)^2 + 106}{2.2 \times 5.1}$

3. Sukvinder decided to decorate her living room. The total area of the walls was 48m². If one roll of wallpaper covers 5m² of wall, how many rolls of wallpaper did Sukvinder need?

4. Thomas earned £109.25 for working a 23-hour week. How much was he paid per hour? Check your calculation by estimating.

Ratios

- **A ratio is used to compare two or more related quantities.**
- **'Compared to' is replaced with two dots : For example, '16 boys compared to 20 girls' can be written as 16 : 20.**
- **To simplify ratios, divide both parts of the ratio by the highest factor. For example, 16 : 20 = 4 : 5 (Divide both sides by 4).**

Examples
- **Simplify the ratio 21 : 28.**

   21 : 28 = 3 : 4
- **The ratio of red flowers to yellow flowers can be written:**

   $10 : 4$

   $= \frac{10}{2} : \frac{4}{2}$

   $= 5 : 2$

   In other words, for every 5 red flowers there are 2 yellow flowers.

   To express the ratio 5 : 2 in the ratio n : 1, divide both sides by 2.

   $= \frac{5}{2} : 2$

   $= 2.5 : 1$

Sharing a quantity in a given ratio

- Add up the total parts of the ratio, e.g. 10:14 totals 14 parts.
- Work out what one part is worth by dividing the quantity by the total number of parts.
- Work out how the quantity is shared by multiplying the values of one part with the ratio values.

Example

£20 000 is shared in the ratio 1 : 4 between Ewan and Leroy.

How much does each receive?

1 + 4 = 5 parts

5 parts = £20 000

1 part = $\frac{£20\,000}{5}$ = £4000

So Ewan gets 1 × £4000 = £4000 and Leroy gets 4 × £4000 = £16 000.

Increasing and decreasing in a given ratio

- Divide the given value by its corresponding ratio value to get one part.
- Multiply the value of one part with the ratio value to get the second value.

Example

A photograph of length 9 cm is to be enlarged in the ratio 5 : 3.

What is the length of the enlarged photograph?

- Divide 9 cm by 3 to get 1 part.

   9 ÷ 3 = 3
- Multiply this by 5. So 5 x 3 = 15 cm on the enlarged photograph.

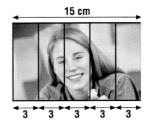

Example

It took 8 people 6 days to build a house.

At the same rate how long would it take 3 people?

Time for 8 people = 6 days.

Time for 1 person = $8 \times 6 = 48$ days.

It takes 1 person longer to build the house.

3 people will take $\frac{1}{3}$ of the time taken by 1 person.

Time for 3 people = $\frac{48}{3} = 16$ days.

Example

A recipe for 4 people needs 1600 g of flour.

How much is needed for 6 people?

- Divide 1600 g by 4, so 400 g for 1 person.
- Multiply by 6, so 6×400 g = 2400 g for 6 people.

Top Tip

When answering problems of the type shown here always try to work out what a unit (or one) is worth. You should then be able to work out what any other value is worth.

Best buys

Use unit amounts to decide which is the better value for money.

Example

The same brand of coffee is sold in two different sized jars.
Which jar represents the better value for money?

- Find the cost per gram for both jars.

  100 g costs 186p so $186 \div 100 = 1.86$p per gram.

  250 g costs 247p so $247 \div 250 = 0.988$p per gram.
- Since the larger jar costs less per gram it is better value for money.

Quick Test

1. Write the following ratios in their simplest form:
   a) 12 : 15
   b) 6 : 12
   c) 25 : 10

2. Three sisters share 60 sweets between them in the ratios 2 : 3 : 7. How many sweets does each sister receive?

3. If 15 oranges cost £1.80, how much will 23 identical oranges cost?

4. A map is being enlarged in the ratio 12 : 7. If the road length was 21 cm on the original map, what is the length of the road on the enlarged map?

Answers 1. a) 4 : 5 **b)** 1 : 2 **c)** 5 : 2 **2.** 10, 15 and 35 sweets respectively **3.** £2.76 **4.** 36 cm

Proportion

Direct proportion and direct proportion graphs

Properties of direct proportion between two quantities

- As one doubles the other doubles.
- As one halves the other halves.
- As one quantity increases the other increases.
- As one quantity decreases the other decreases.

This is what a direct proportion graph looks like:

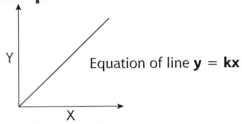

Equation of line $y = kx$

So **y is proportional to x**, and we write $y \propto x$, or $y = kx$, where k is a constant.
($\propto$ means proportional to)

Direct proportion formulae

Whenever you read '*y is proportional to x*' or '*y varies with x*' in a question, you can write down:

$$y \propto x, \text{ then } y = kx \text{ (k is a constant)}$$

In most proportion questions you have to find a value for k, then write a formula that connects y and x.

Example

y is proportional to x. When y is 36 x is 9:

a. Find a formula connecting y and x when y is 36 and x is 9.

$y \propto x$
so $y = kx$
$36 = k \times 9$ so $k = 4$
The formula is $y = 4x$

b. If $y = 4x$, then calculate y when $x = 12$
$y = 4x = 4 \times 12 = 48$

Example

Sarah's pay (P) varies directly as the number of hours (H) she has worked.

Her pay is £63 when she works 9 hours.

a. Find a formula for P.

$P \propto H$
$P = kH$
$63 = k \times 9$ so $k = 7$
formula $P = 7H$

b. Calculate her pay for working 30 hours
$P = 7 \times 30 = £210$

Direct proportion: $y = kx^2$ and $y = k\sqrt{x}$

y and x are still proportional here but we can only use these rules:
- As one quantity increases the other increases.
- As one quantity decreases the other decreases.

Example

y varies directly as x^2, so $y = kx^2$

When $y = 45$, $x = 3$

a. Find the formula connecting y and x.

$y = kx^2$

$45 = k \times 32$

$45 = k \times 9$

$k = 5$ so the formula is $y = 5x^2$

b. Calculate y when $x = 4$

$y = 5 \times 4^2$

$y = 5 \times 16 = 80$

Example

a varies directly as $\sqrt{b}$, so $a = k\sqrt{b}$

When $a = 35$, $b = 25$

a. Find a formula connecting a and b.

$35 = k \times \sqrt{25}$

$35 = k \times 5$

$k = 7$ so the formula is $a = 7\sqrt{b}$

b. Calculate a when $b = 144$

$a = 7 \times \sqrt{144}$

$a = 7 \times 12 = 84$

Inverse proportion

The following rules are applied **if y is inversely proportional to x:**
- If y doubles x halves.
- If y halves x doubles.

Formula for inverse proportion $y = \dfrac{k}{x}$

so $k = yx$

Example

y varies inversely to x. When $y = 3$, $x = 20$.

a. Find the formula connecting x and y

$3 = \dfrac{k}{20}$

$k = 3 \times 20 = 60$ so the formula is $y = \dfrac{60}{x}$

b. Find y when $x = 4$

$y = \dfrac{60}{4} = 15$

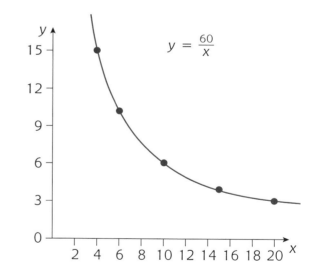

$$y = \frac{60}{x}$$

Quick Test

1. f varies with g. When $f = 12$, $g = 2.4$.
  a. Write a formula connecting f with g.
  b. What is f when $g = 14$

2. t varies inversely with v. Find a formula connecting t and v when $t = 6$ and $v = 8$.

3. The area (A cm) of a circle varies directly with the square of its radius (r cm).
  a. If $A = 706.5$ cm and $r = 15$ cm, find the formula connecting A and r.
  b. What is A when $r = 25$ cm

Money

Wages and salaries

This is a very useful topic as it is important that people can work out how much they are earning to make sure they are receiving the correct amount.

Example

Anna gets paid £18720 per annum. What is her weekly wage?

£18720 ÷ 52 = £360 per week

Example

As a bus driver, Joe earns £12.60 per hour. How much will he earn after working 25 hours?

25 × £12.60 = £315

Overtime

In some jobs the rate of pay is higher for people working at night, weekends or holidays.

- Double time is the normal rate of pay × 2.
- Time and a half is the normal rate of pay × 1.5.

Example

Mary gets paid £5.50 per hour. If she works on a Saturday she gets double time.

a. Calculate Mary's pay for 23 hours of weekday work.
   £5.50 × 23 = £126.50
b. Calculate Mary's pay for 8 hours on Saturday.
   (£5.50 × 2) × 8 = £88

Commission

In many sales jobs the employee's salary is dependent on the amount they sell. This is commission. A good sales person will make a lot of commission.
In some jobs employees earn a small set wage and commission.

Example

Colin sells kitchens. He earns 13% commission on each kitchen he sells.

a. How much is he paid for selling a £2000 kitchen?
   13% of £2000 = 0.13 × £2000 = £260 commission
b. In the summer he sold £35000 worth of kitchens. What was his commission for the summer?
   13% of £35000 = 0.13 × £35000 = £4550 commission

Example

Samantha works in a hi-fi shop. She earns a monthly salary of £400 and 8% commission on her sales.

a. In June she sold £4000 worth of stock. What was her commission?
   8% of £4000 = $\frac{8}{100}$ × £4000 = £320
b. In December Samantha earned £2300. What was her commission?
   £2300 – £400 = £1900

Hire purchase

Hire purchase is a way of paying for a product over a period of time.
This is useful as people can get a product and pay it off over time.

How hire purchase works

- A deposit is made and the product can be taken by the customer.
- The customer pays weekly or monthly instalments until the product is fully paid for.
- When an item is bought through hire purchase, it usually ends up costing more than it would have if the product was paid for outright.

Example

The cash price for a couch is £1100. To pay for the couch through hire purchase a 15% deposit has to be paid then twelve monthly instalments of £90.

a. How much will the deposit be?
15% of £1100 = 0.15 × £1100 = £165

b. How much would be paid for all 12 of the instalments?
12 × £90 = £1080

c. What is the hire purchase price of the couch?
deposit + instalments = £165 + £1080 = £1245

d. How much more is this than the cash price?
hire purchase price – cash price = £1245 – £1100 = £145

Foreign exchange

The rate of exchange for each currency will normally be given by an amount per £.

Exchange rates:
£1 = €1.65 (euros) – exchange rate is 1.65
£1 = $1.85 (US Dollars) – exchange rate is 1.85

Important rules
Number of Pounds × Exchange Rate = Foreign Money
Foreign Money ÷ Exchange Rate = Number of Pounds

Example

Tina saves up her pocket money for a family holiday in the US. She takes her savings of £85 to the bank. How many dollars will she get in return?

85 × 1.85 = $157.25

Example

Colin spends the summer working in Spain. When he returns he exchanges 462 euros into pounds.
His bank charges 2% commission. How much money will Colin get?

462 ÷ 1.65 = £280

The bank takes 2% of this as commission

2% of £280 = £5.60

£280 – £5.60 = £274.40   This is the amount that Colin will get.

Quick Test

1. Cheryl gets £580 in her weekly wage packet for working 40 hours. What is her hourly rate of pay?

2. Andrew buys a car through hire purchase. He pays a deposit of £1500 and 36 installments of £125. How much does Andrew pay for the car?

3. Robert returns from a school trip in Germany with €85. Use the exchange rate above to find out how many pounds he will get back.

Speed, distance and time

Distance-time graphs

Distance-time graphs are a good way of representing a journey.

You have to know how to interpret them. Here are things to look out for:

- When the line is steep, this means that vehicle is going fast.
- When the line is less steep, the vehicle is slower.
- When the line is horizontal, the object is not moving.

Example

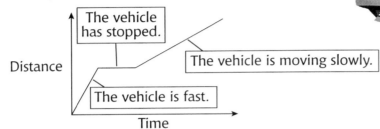

The vehicle has stopped.

The vehicle is moving slowly.

Distance

The vehicle is fast.

Time

Speed

Speed tells us how fast an object is moving.

To calculate the average speed of an object we need to know two things about it:

1. the distance it travels; and

2. the time it takes to travel this distance.

The formula to find average speed $S = \frac{D}{T}$

Example

Calculate the average speed in miles per hour (mph) of a plane flying from Paris to Edinburgh.

The distance is 620 miles and it takes 2 hours.

$S = \frac{620}{2} = 310 \, \text{mph}$

Formulae to remember

Speed: $S = \frac{D}{T}$

Distance: $D = S \times T$

Time: $T = \frac{D}{S}$

T = Time S = Speed D = Distance

Distance

The formula to find the distance is $D = S \times T$

The units for distance can range from centimetres to miles.

Example

Samera jogs at an average speed of 6km/h for 3 hours. What distance does she jog?

$D = S \times T \quad S = 6\text{km} \quad T = 3$

$D = 6 \times 3 = 18 \, \text{km}$

Example

It takes the Owen family two-and-a-half hours driving at a speed of 70 mph to drive to a theme park. What is the distance the Owen family has to drive to get to the theme park?

$D = S \times T \quad S = 70 \text{ miles} \quad T = 2.5$

$D = 70 \times 2.5 = 175 \text{ miles.}$

Time

The formula to find time is $T = \frac{D}{S}$

Example

Use the formula to calculate the time for each journey.

a. Sarfraz walks 20 miles at a speed of 4 mph

$T = \frac{20}{4} = 5$ hours

b. Amanda drives 270 km at a speed of 60 km/h.

$T = \frac{270}{60} = 4.5$ hours

Units

When doing speed, distance and time questions it is important that the units correspond.

- In a question to find distance the units for time have to match up with the units of time for the speed. For example, if the speed is in km/h and the time is in minutes, you will get the wrong answer unless you change the unit of time into hours.

- If you have a question where the unit of distance does not match up with the unit of distance in the speed, e.g. if the distance is in metres and the speed is in km/h, then you will get the wrong answer unless you make the units both in metres or both in kilometres.

> Conversion of minutes to hours and hours to minutes
> **Formula** $\frac{\text{Number of minutes}}{60}$ = time in hours
> **Formula** Number of hours × 60 = time in minutes

Example

Leo walks for 42 minutes at an average speed of 6 km/h. What is the distance that Leo walks?

First of all we have to convert the time into hours: $\frac{42 \text{ minutes}}{60 \text{ minutes}} = 0.7$ hours

$D = S \times T = 6 \times 0.7 = 4.2$ km

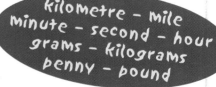

Example

In a 1500m race the winner's average speed is 15 km/h.

a. Find the time in hours that it takes for the winner to run the race.
Make the unit for the distances the same: 1500 ÷ 1000 = 1.5 km

$T = \frac{D}{S} = \frac{1.5}{15} = \frac{1}{10}$ hour

b. What is the runner's time in minutes? 0.1 × 60 = 6 minutes

Quick Test

1. Calculate the average speed of a journey of 88km which takes 4 hours.

2. Calculate the distance traveled in 3 hours and 30 minutes at an average speed of 64km/h

3. Mr. Smith left his house at 9:30 am and drives 260 miles to Glasgow at an average speed of 50 mph. When does he arrive?

Answers 1. 22km/h **2.** 224 km **3.** 2 :42pm

Standard index form

Standard index form is used to write very large numbers or very small numbers in a simpler way. When written in standard form the number will be written as:

$$a \times 10^n$$

a must be between 1 and 10, $1 \leq a < 10$
The value of *n* is the number of times the decimal point has to move to the right to return the number to its original value.

Learn these three rules:
1) The front number must always be between 1 and 10.
2) The power of 10, *n*, is purely how far the decimal point moves.
3) If the number is **BIG**   If the number is **SMALL**
   n is positive.   *n* is negative.

Big numbers

Examples
- Write 6 230 000 in standard form.
  Move the decimal point to between the 6 and 2 to give 6.230000 ($1 \leq 6.23 < 10$).
  Count how many places the decimal point needs to be moved to restore the number.
  6 230 000   (6 places)
  In standard form 6 230 000 = 6.23×10^6

- 4371 = 4.371×10^3 in standard form.

Small numbers

Examples
- Write 0.003 71 in standard form.
  Move the decimal point to between the 3 and 7 to give 3.71 ($1 \leq 3.71 < 10$).
  Count how many places the decimal point has been moved.
  0.003 71   (3 places)
  In standard form 0.00371 = 3.71×10^{-3}

A negative number means that the decimal point is moved to the left.

0.000 047 9 = 4.79×10^{-5} in standard form.

Watch out!

Several common mistakes when answering standard form questions are:
- Reading a calculator display incorrectly and writing down 2.4^7 instead of 2.4×10^7.
- Forgetting to write the answer in standard form particularly on the non-calculator paper.

  e.g. $(2 \times 10^6) \times (6 \times 10^3) = (2 \times 6) \times (10^6 \times 10^3)$
  $$= 12 \times 10^9$$
  $$= 1.2 \times 10^{10}$$

Standard form with the calculator

To key a number in standard form into the calculator, use the **EXP** key. (Some calculators use **EE**. Make sure that you check your calculator, as calculators vary greatly.)

Examples

6.23×10^6 can be keyed in as: `6` `.` `2` `3` `EXP` `6`

4.93×10^{-5} can be keyed in as: `4` `.` `9` `3` `EXP` `5` `+/-`

Most calculators do not show standard form correctly on the display.

`7.632 09` means 7.632×10^9. `4.62 -07` means 4.62×10^{-7}.

Remember to put in the $\times 10$ part if it has been left out.

Calculations with standard form

Use the calculator to do complex calculations in standard form.

Examples

$(2.6 \times 10^3) \times (8.9 \times 10^{12}) = 2.314 \times 10^{16}$

This would be keyed in as:

`2` `.` `6` `EXP` `3` `×` `8` `.` `9` `EXP` `1` `2` `=`

Just key in as usual:

`2` `.` `7` `EXP` `3` `+/-`

Check that for $(1.8 \times 10^6) \div (2.7 \times 10^{-3})$ the answer is 6.6×10^8

If a calculation with standard form is on the non-calculator paper, **the laws of indices can be used when multiplying and dividing numbers written in standard form.**

Examples

$(2.4 \times 10^{-4}) \times (3 \times 10^7)$

$= (2.4 \times 3) \times (10^{-4} \times 10^7)$

$= 7.2 \times (10^{-4+7})$

$= 7.2 \times 10^3$

$(12.4 \times 10^{-4}) \div (4 \times 10^7)$

$= (12.4 \div 4) \times (10^{-4} \div 10^7)$

$= 3.1 \times (10^{-4-7})$

$= 3.1 \times 10^{-11}$

Top Tip
Standard form questions are very common on both the calculator and non-calculator papers.

Quick Test

1. Write the following numbers in standard form:
   a) 630 000
   b) 2730
   c) 0.000 042 9
   d) 0.000 000 63

2. Without a calculator, work out the following, leaving your answer in standard form.
   a) $(2 \times 10^5) \times (3 \times 10^7)$
   b) $(6.1 \times 10^{12}) \times (2 \times 10^{-4})$
   c) $(8 \times 10^9) \div (2 \times 10^6)$
   d) $(6 \times 10^8) \div (2 \times 10^{-10})$

3. Work these out on a calculator. Give your answers to 3 s.f.
   a) 1.279×10^9
   b) $\dfrac{(1.693 \times 10^4) \times (2.71 \times 10^{12})}{2.94 \times 10^{-2}}$

4. Calculate, giving your answers in standard form correct to 3 s.f.

   $\dfrac{(3.72 \times 10^8) - (1.6 \times 10^4)}{3.81 \times 10^{-3}}$

Test your progress

Use the questions to test your progress.
Check your answers at the back of the book on page 94.

1. From this list of numbers (2, 9, 21, 40, 41, 64, 100):
a) Write down the numbers that are odd ....
b) Write down the square numbers ....
c) Write down the prime numbers ....
d) Write down any numbers which are factors of 80 ....
e) Write down any numbers which are multiples of 4 ....

2. The temperature outside is −6°, inside it is 15° warmer.
What is the temperature inside? ....

3. Work out the answers to:
a) $589 \times$
$ 76$

b) $62\overline{)1674}$

....

4. Mohammed needs some tiles for the kitchen floor. The tiles are sold in boxes of 8.
Mohammed works out that he needs 60 tiles. How many boxes will he need to buy?

....

5. Hussain scored 58 marks out of 75 in a test. What percentage did he get?

....

6. A school raises £525 at the summer fair. 60% of the money raised is used to repair the tennis courts.
How much is used to repair the tennis courts? ....

7. The ingredients for 8 small cakes are:
• 300 g self-raising flour
• 150 g butter
• 250 g sugar
• 2 eggs
Andrew is making 20 small cakes. Write down the amounts of ingredients he will need.
.... g self-raising flour
.... g butter
.... g sugar
.... eggs

8. Mrs Patel inherits £55 000. She divides the money between her children in the ratio 3 : 3 : 5.
How much does the child with the largest share receive?

....

9. Work these out on your calculator, giving your answers to 3 s.f.
a) $\dfrac{4.2\,(3.6 + 5.1)}{2 - 1.9} =$

b) $\dfrac{3.8 + 4.6}{2.9 \times 4.1} =$

....

10. A cyclist travels for 3 hours and 15 minutes at a speed of 32 km/h. What distance does he travel?

11. Toothpaste is sold in three different sized tubes.
50 ml = £1.24
75 ml = £1.96
100 ml = £2.42
Which of the tubes of toothpaste is the best value for money?
You must show full working out in order to justify your answer.
....
....

12. Find a formula for each of the following:
a) y varies with x. $y = 15$, $x = 2$ ....
b) m is proportional to n^2. $m = 32$, $n = 2$ ....
c) h is inversely proportional to j. $h = 3.5$, $j = 7$ ....

13. William gets 5% commission as an estate agent. In July he sold houses totalling a value of £250 000. Work out his commission.
....

14. Amanda works in a pharmacy. She gets paid £6.20 per hour. She works 20 hours during the week and gets time-and-a-half for working 5 hours on a Sunday. How much is Amanda paid for that week?
....

15. James put £632 in a new savings account. The interest is 2.5% p.a. How much interest will he get after 4 months?
....

16. The price of a television has risen from £350 to £420. Work out the percentage increase in the price.
....
....

17. Write these numbers in standard form:
a) 2 670 000
....
b) 4270
....
c) 0.032 96
....
d) 0.027
....

18. The mass of a hair is 0.000 042 g.
a) Write this number in standard form ....
b) Calculate, in standard form, the mass of 63 105 hairs ....

19. Jacky travels to New Zealand with £600. The exchange rate is £1 = $3.55.
When she converts her money to New Zealand dollars, how much will she get?
....

How did you do?

1–5	correct	....	start again
6–10	correct	....	getting there
11–15	correct	....	good work
16–19	correct	....	excellent

Algebra 1

Algebraic conventions

- A **term** is a collection of numbers, letters and brackets, all multiplied together.
- Terms are separated by + and − signs. Each term has a + or − **attached to the front of it**.

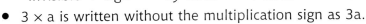

$$3xy - 5r - 2x^2 + 4$$

invisible + sign xy term r term x^2 term   number term

- 3 × a is written without the multiplication sign as 3a.

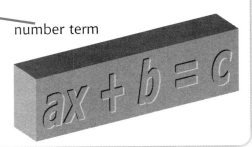

a + a + a = 3a   2ab = 2ba = a2b = ba2 etc.

$a \times a \times a = a^3$, not 3a

$a \times a \times 2 = 2a^2$, not $(2a)^2$

$a \times b \times 2 = 2ab$

Collecting like terms

Expressions can be simplified by collecting **like terms**.

Only collect the terms if their letters and powers are **identical**.

Examples

4a + 2a = 6a

$3a^2 + 6a^2 - 4a^2 = 5a^2$

4a + 6b − 3a + 2b = a + 8b

9a + 4b cannot be simplified since there are no like terms.

3xy + 2yx = 5xy

> Add the a terms together, then the terms with b. Remember a means 1a.

> Remember xy means the same as yx

Writing formulae

Quite often in the exam you are asked to write a formula when given some information or a diagram.

Example

Frances buys x books at £2.50 each. She pays with a £20 note.
If she receives £C change write down as a formula.

> notice that no £ signs are put in our formula

$$C = 20 - 2.50x$$

> this is the amount of money she spent

If in doubt, check by substituting a value for x, i.e. if she bought 1 book $x = 1$ so her change would be 20 − 2.50 × 1 = £17.50

Example

Some patterns are made by using grey and white paving slabs.

Write a formula for the number of grey paving slabs (g) in a pattern that uses (w) white ones.

Formula is g = 2w + 2

2w represents the 2 layers, + 2 gives the grey slabs which are on either end of the white ones.

Formulae, expressions and substituting

p + 3 is an **expression**.

y = p + 3 is a **formula**. The value of y depends on the value of p.

Replacing a letter with a number is called substitution. When substituting:

- Write out the expression first and then replace the letters with the values given.
- Work out the value on your calculator. Use bracket keys where possible and pay attention to the order of operations.

Try these out on your calculator

Examples

Using W = 5.6, t = −7.1 and u = $\frac{2}{5}$, find the value of these expressions, giving your answers to 3 s.f.

a) $\frac{W + t}{u}$ **b)** $W - \frac{t}{u}$ **c)** $\sqrt{Wt^2}$

Remember to show the substitution.

a) $\frac{W + t}{u} = \frac{5.6 + (-7.1)}{\frac{2}{5}} = -3.75$

b) $W - \frac{t}{u} = 5.6 - \frac{(-7.1)}{\frac{2}{5}} = 23.4$

c) $\sqrt{Wt^2} = \sqrt{5.6 \times (-7.1)^2} = 16.81$

You may need to treat t^2 as $(-7.1)^2$, depending on your calculator.

Using formulae

A formula describes the relationship between two (or more) variables.

A formula must have an = sign in it.

Example

Andrew hires a van. There is a standing charge of £8 and then it costs £3 per hour. How much does it cost for:

a) 6 hours' drive?

b) y hours' drive?

c) Write a formula for the total hire cost C.

a) 8 + (3 × 6) = £26

b) 8 + (3 × y) = 8 + 3y

d) C = 8 + 3y

This is a formula which works out the cost of hiring the van for any number of hours.

Top Tip
When substituting in to an expression or formula you must show each step in your working out. By showing your substitution you will obtain method marks even if you get the final answer wrong.

Quick Test

1. Simplify these expressions by collecting like terms:
   a) 5a + 2a + 3a
   b) 6a − 3b + 4b + 2a
   c) 5x − 3x + 7x − 2y + 6y
   d) $3xy^2 - 2x^2y + 6x^2y - 8xy^2$

2. Using p = 6.2, r = −3.2 and s = $\frac{2}{3}$, find the value of these expressions, giving your answer to 3 s.f.
   a) pr + s   **b)** $p^2s - r$
   c) $r^2 - p/s$ **d)** $(ps - r)^2$

Algebra 2

Multiplying out brackets

- This helps to simplify algebraic expressions.
- The term **outside** the brackets **multiplies each separate term inside the brackets**.

Examples

$3(2x + 5) = 6x + 15$ $(3 \times 2x = 6x, 3 \times 5 = 15)$

$a(3a - 4) = 3a^2 - 4a$ $b(2a + 3b - c) = 2ab + 3b^2 - bc$

If the term outside the bracket is **negative**, all of the signs of the terms inside the bracket are **changed** when multiplying out.

Examples

$-4(2x + 3) = -8x - 12$ $-2(4 - 3x) = -8 + 6x$

To simplify expressions, expand the brackets first then collect like terms.

Examples

Expand and simplify $2(x - 3) + x(x + 4)$.

$2(x - 3) + x(x + 4)$

$= 2x - 6 + x^2 + 4x$ Multiply out the brackets.

$= x^2 + 6x - 6$ Collect like terms.

> The subject of the formula is usually written first.

> **Top Tip**
> If you are asked to expand brackets it just means multiply them out. When you have finished multiplying out the brackets simplify by collecting like terms in order to pick up the final mark.

Factorisation (putting brackets in)

This is the reverse of expanding brackets.
An expression is put into brackets by taking out **common factors**.

One bracket

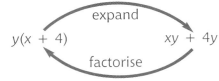

$y(x + 4)$ →expand→ $xy + 4y$ ←factorise←

To factorise $xy + 4y$:

- recognise that y is a factor of each term;
- take out the common factor;
- the expression is completed inside the bracket, so that the result is equivalent to $xy + 4y$, when multiplied out.

Examples

Factorise the following.

a) $5x^2 + x = x(5x + 1)$

b) $4x^2 + 8x = 4x(x + 2)$

c) $5x^3 + 15x^4 = 5x^3(1 + 3x)$

Factorising can be useful when simplifying algebraic fractions.

Example

Simplify $\frac{5x + 15}{(x + 3)} = \frac{5\cancel{(x + 3)}}{\cancel{(x + 3)}} = 5$

Quick Test

1. Multiply out the brackets and simplify where possible:
   - **a.** $3(x + 2)$
   - **b.** $2(x + y)$
   - **c.** $-3(x - 4)$
   - **d.** $4(x + 7) + 2x$
   - **e.** $3(4x - 2) + 2(x + 7)$
   - **f.** $-2(5 - 2x) - (4 + x)$

2. Factorise the following expressions:
   - **a.** $3x + 6$
   - **b.** $10y - 15$
   - **c.** $24 - 16m$
   - **d.** $x^2 - 4x$
   - **e.** $2xy + 8y$
   - **f.** $20c - 2c^2$

Answers 1a. $3x + 6$ b. $2x + 2y$ c. $-3x + 12$ d. $6x + 28$ e. $14x + 8$ f. $3x - 14$ 2a. $3(x + 2)$ b. $5(2y - 3)$ c. $8(3 - 2m)$ d. $x(x - 4)$ e. $2y(x + 4)$ f. $2c(10 - c)$

Number patterns and sequences

A sequence is a list of numbers. There is usually a relationship between the numbers. Each value in the list is called a term.

Example
The odd numbers form a sequence 1, 3, 5, 7, 9, 11, . . . in which the terms have a common difference of 2.

The common difference is 2.

Important number sequences

Square numbers	1, 4, 9, 16, 25, . . .
Cube numbers	1, 8, 27, 64, 125, . . .
Triangular numbers	1, 3, 6, 10, 15, . . .
The Fibonacci sequence	1, 1, 2, 3, 5, 8, 13, . . .

Sequence

The nth term of a sequence is often denoted by N.

$$N = an + b$$

Examples

1. For the sequence of odd numbers, find an expression for the nth term.
   1, 3, 5, 7, 9, ...
   - Find the common difference, 'a'.
     a = 2 here. So N = 2n + b
   - Now substitute the value of the first term
     n = 1
     1 = 2 + b
     So b = −1
   - Formula for nth term N = 2n −1
   - Check for the 10th term
     So n = 10
     N = 20 − 1 = 19

2. For the sequence 7, 10, 13, 16, 19 ... find an expression for the nth term.
   - Find the common difference − 'a'.
     So a = 3 here. So N = 3n + b
   - Now substitute the value of the first term
     N = 7 when n = 1
     7 = 3 + b
     So b = 4
   - Formula for nth term N = 3n + 4
   - Check for the 10th term
     So n = 10
     N = 30 + 4 = 34

Quick Test

1. Write down the next two terms in the sequences below:
   a. 5, 7, 9, 11, __, __
   b. 3, 10, 17, 24, __, __
   c. 1, 4, 9, 16 , __, __

2. Write down the nth term in the sequences below:
   a. 5, 7, 9, 11 ...
   b. 2, 5, 8, 11 ...
   c. 6, 10, 14, 18 ...

Answers 1. a 13, 15 b. 31, 38 c. 25, 36 2. a. N = 2n + 3 b. N = 3n − 1 c. N = 4n + 2

Equations

Solving linear equations of the form ax + b = c

When solving equations, the balance method is used; whatever is done to one side of the equation must be done to the other.

Example

Solve $2x + 15 = 9$
       $2x = 9 - 15$ Subtract 15 from both sides.
       $2x = -6$
       $x = -6 \div 2$ Divide both sides by 2.
       $x = -3$

Example

Solve $3x - 18 = 6$
       $3x = 6 + 18$
       $3x = 24$
       $x = 24 \div 3$
       $x = 8$

Top Tip

Solving equations is a very common topic in Standard Grade. Try to work through them in a logical way, always showing full working out. If you have time, check by substituting your answer back into the equation to see if it works.

Solving linear equations of the form ax + b = cx + d

The trick with this type of equation is to get the xs together on one side of the equal sign and the numbers on the other side.

Example

Solve: $5x - 9 = 12 - 4x$
       $9x - 9 = 12$ Add 4x to both sides.
       $9x = 21$ Add 9 to both sides.
       $x = \frac{21}{9} = 2\frac{1}{3}$

Top Tip

If in the exam you do not know that $\frac{21}{9} = 2\frac{1}{3}$, leave it as $\frac{21}{9}$ to obtain full marks!

Example

Solve $10x + 8 = 2x + 40$
       $8x + 8 = 40$ subtract 2x from both sides
       $8x = 32$ subtract 8 from both sides
       $x = 32 \div 8$
       $x = 4$

Solving linear equations with brackets

Don't be put off if an equation has brackets.

It's just the same as the other equations once the brackets have been multiplied out.

$3(x - 2)$
$= 2(x + 6)$

Examples

a) Solve:

$3(x - 2) = 2(x + 6)$

$3x - 6 = 2x + 12$ Multiply out the brackets first.

$x - 6 = 12$

$x = 18$

b) Solve:

$5(x - 2) + 6 = 3(x - 4) + 10$

$5x - 10 + 6 = 3x - 12 + 10$

$5x - 4 = 3x - 2$

$2x = 2$

$x = 1$

Using equations to solve problems

Example

The perimeter of the triangle is 20 cm.

Work out the value of x and hence find the length of the 3 sides.

$x + 2x + 5 + 4x + 1 = 20$

$7x + 6 = 20$

$7x = 20 - 6$

$7x = 14$

$x = \frac{14}{7}$

$x = 2$

Triangle with sides labelled x, $2x + 5$, and $4x + 1$.

The perimeter is found by adding lengths together. Collect like terms. Solve the equation as before.

So the lengths of the sides are 2 (= x), 9 (= 4x + 1) and 9 (= 2x + 5)

Quick Test

Solve the following equations:

1. $5x - 2 = 13$

2. $4x + 11 = 51$

3. $7(x + 2) = 14$

4. $5x + 3 = 2x + 9$

5. $6x - 1 = 15 + 2x$

6. $3(x + 2) = x + 4$

7. $3(x - 3) = 2(x - 3)$

Inequalities

Inequalities are solved in a similar way to equations.
Multiplying and dividing by negative numbers changes the direction of the sign.
For example if $-x \geq 5$ then $x \leq -5$.

The four inequality symbols

> means 'greater than' $\geq$ means 'greater than or equal to'
< means 'less than' $\leq$ means 'less than or equal to'

Use ● when the end point is included and ○ when the end point is not included on a number line.

So $x > 3$ and $3 < x$ both say 'x is greater than 3'

Examples

Solve the following inequalities:

a) $4x - 2 < 2x + 6$
  $2x - 2 < 6$ Subtract 2x from both sides.
  $2x < 8$ Add 2 to both sides.
  $x < 4$ Divide both sides by 2.

The solution of the inequality may be represented on a number line.

4

b) $-5 < 3x + 1 \leq 13$ Subtract 1 from each part.
  $-6 < 3x \leq 12$ Divide by 3.
  $-2 < x \leq 4$

The integer values which satisfy the above inequality are

$-1, 0, 1, 2, 3, 4$.

Graphs of inequalities

The graph of an equation such as $y = 3$ is a line, whereas the graph of the inequality $y < 3$ is a region which has the line $y = 3$ as its **boundary**.

To show the region for given inequalities:
- Draw the boundary lines first.
- For **strict** inequalities > and <, the boundary line is not included and is shown as a dotted line.
- It is often easier with several inequalities to shade out the unwanted regions, so that the solution is shown **unshaded**.

Example

The diagram shows unshaded the region $x > 1$, $x + y \leq 4$, $y \geq 0$

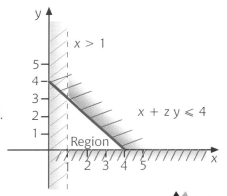

$x > 1$

$x + z\,y \leqslant 4$

Region

Quick Test

Solve the following inequalities:

1. $2x - 3 < 9$

2. $5x + 1 \geq 21$

3. $1 \leq 3x - 2 \leq 7$

4. $1 \leq 5x + 2 < 12$

Top Tip
Inequalities are solved in a similar way to equations.

Sample exam-style questions

1. a. Multiply out the brackets and simplify: $5x + 3(2x - 4)$

　　　　　　　　　Multiply out: $5x + 6x - 8$

　　　　　　　　　　Simplify: $11x - 8$

b. 　　　Solve algebraically the equation: $6x - 4 = 4x - 8$

　　　Subtract 4x from both sides: $2x - 4 = -8$

　　　　Add 4 to both sides: $2x = -4$

　　　Divide both sides by 2: $x = -2$

2.

　Pattern 1　　　　Pattern 2　　　　Pattern 3　　　　Pattern 4

Pattern (x)	1	2	3	4	5
Number of lines (y)	4	7	10	13	

a. How many lines will there be in pattern 5?

3 lines are being added from one pattern to the next so

$13 + 3 = 16$

There will be 16 lines in pattern 5.

b. Write down a formula for the number of lines, y, when you know the pattern number, x.

$y = ax + b$

The common difference is 3, so a = 3 here.

So $y = 3x + b$

We know that when x = 1 y = 4, so substitute the values into the formula $4 = 3 + b$

So b = 1

Formula $y = 3x + 1$

c. What pattern will have 43 lines?

Here y = 43, so substitute it into the formula.

$43 = 3x + 1$

$43 - 1 = 3x$

$42 = 3x$

$42 \div 3 = x$

$x = 14$

The 14th pattern will have 43 lines in it.

3. a. Factorise $12x - 48$

12 is the highest common factor, so this will go outside the brackets

$12(x - 4)$

b. Solve algebraically the inequality

　　　　　　　　　$4x + 8 \leq 20$

Subtract 8 from both sides　　$4x \leq 12$

Divide each side by 4　　　　$x \leq 12 \div 4$

　　　　　　　　　$x \leq 3$

Straight line graphs

Drawing straight line Graphs

In order to draw a straight line graph follow these easy steps:-

Step 1
Choose 3 values of x and draw up a table.

Step 2
Work out the value of y for each value of x.

Step 3
Plot the coordinates and join up the points with a straight line.

Step 4
Label the graph.

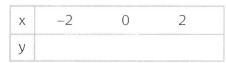

Example

Draw the graph of $y = 3x - 1$

1. Draw up a table with some suitable values of x.

x	−2	0	2
y			

2. Work out the y values by putting each x value into the equation.

$$y = 3x - 1$$

e.g. if $x = -2$

$$y = (3 \times -2) - 1$$
$$y = -6 - 1$$
$$y = -7$$

3. Compete the table with the y values.

x	−2	0	2
y	−7	−1	5

4. Plot the points and draw the line.

Graphs of $y = a$, $x = b$

$y = a$ is a horizontal line with every y coordinate equal to a.

$x = b$ is a vertical line with every x coordinate equal to b.

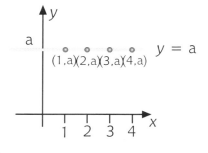

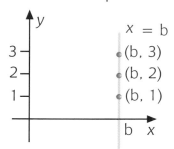

Interpreting $y = mx + c$

The general equation of a straight line graph is $y = mx + c$.

m is the **gradient** (steepness) of the line.

- As m increases the line gets steeper.
- If m is **positive** the line slopes **forwards**.
- If m is **negative** the line slopes **backwards**.
- **Parallel** lines have the **same gradient**.

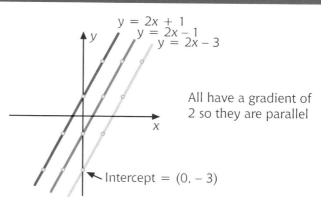

$y = 2x + 1$
$y = 2x - 1$
$y = 2x - 3$

All have a gradient of 2 so they are parallel

Intercept = (0, – 3)

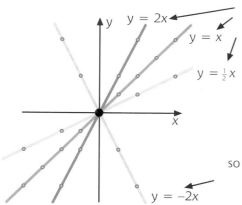

$y = 2x$
$y = x$
$y = \frac{1}{2}x$

As m increases the line gets steeper.

$y = -2x$

m is negative so the graph slopes 'backwards'.

c is the **intercept** on the y axis, that is, where the graph cuts the y axis.

Top Tip
You need to be able to sketch a straight line graph from its equation. If you can do this then you will be able to tell if the graph you have drawn is correct.

Finding the gradient of a line

- To find the gradient, choose two points.
- Draw a triangle as shown.
- Find the change in y (height) and the change in x (base).
- gradient = $\frac{\text{change in } y}{\text{change in } x}$ or $\frac{\text{height}}{\text{base}} = \frac{4}{3} = 1\frac{1}{3}$
- Decide if the gradient is positive or negative.

Do not count the squares as the scales may be different on each axis.

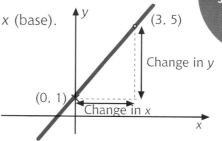

(3, 5)

Change in y

(0, 1)

Change in x

Quick Test

1. Draw the graph of $y = 6 - 2x$. From your graph write down the solution of the equations:
   a) $6 - 2x = 4$
   b) $6 - 2x = 3$

2. Write down the gradient and intercept for each of these straight line graphs:
   a) $y = 4 + 2x$
   b) $y = 3x - 2$
   c) $2y = 6x + 4$

Interpreting graphs

Graphs in practical situations

Linear graphs are often used to show **relationships**.

Example

The graph shows the prices charged by a van hire firm.

- Point A shows how much was charged for hiring the van, i.e. £50.
- The gradient = 20.
  This means that £20 was charged per day for the hire of the van.
  Hence for 5 days' hire the van cost
  £50 + £20 × 5 = £150

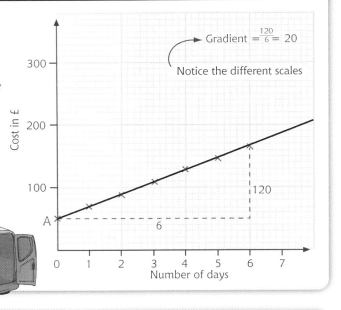

Gradient $= \frac{120}{6} = 20$

Notice the different scales

Cost in £

120

6

Number of days

Distance–time graphs

These are often called **travel graphs**.

The **speed** of an object can be found by finding the gradient of the line.

$$\text{speed} = \frac{\text{distance travelled}}{\text{time taken}}$$

Example

The graph shows Mr Rogers' car journey. Work out the speed of each stage.

a) The car is travelling at 30 mph for 1 hour (30 ÷ 1).

b) The car is stationary for 30 minutes.

c) The graph is steeper so the car is travelling faster, at a speed of 60 mph for 30 minutes (30 ÷ 0.5).

d) The car is stationary for 1 hour.

e) The return journey is at a speed of 40 mph (60 ÷ 1.5).

Distance from home (miles)

Mr Rogers' car journey

30

0.5hr

(a) (b) (c) (d) (e)

0900 1000 1100 1200 1300 Time

Top Tip
Notice the importance of using the gradient of a line. It is useful to note that on the distance–time graph example, the scales on both axes are different. Care must be taken when reading the scales: always make sure you understand the scales before you start.

Conversion graphs

These are used to convert values of one quantity to another, e.g. litres to pints, km to miles, £ to $, etc.

Example

Suppose £1 is worth $1.50.
Draw a conversion graph.

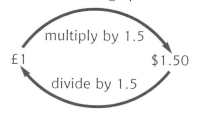

£	1	2	3	4	5
$	1.5	3.0	4.5	6.0	7.5

$\times 1.5$

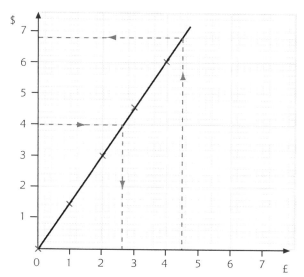

- Make a table of values.
- Plot each of these points on the graph paper.
- To change $ to £, read across to the line then down, e.g. $4 is £2.67 (approx.)
- To change £ to $, read up to the line then read across, e.g. £4.50 is $6.80 (approx.)

Quick Test

1.

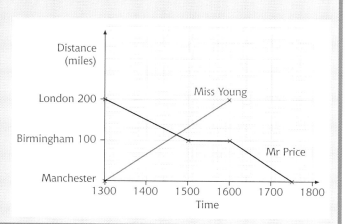

These containers are being filled with a liquid at a rate of 150 ml per second.
The graphs show how the depth of the water changes with time.

Match the containers with the graphs.

2. The travel graph shows the car journeys of two people. From the travel graph find:
   a) the speed at which Miss Young is travelling.
   b) the length of time Mr Price has a break
   c) the speed of Mr Price from London to Birmingham.
   d) the time at which Miss Young and Mr Price pass each other.

Test your progress

Use the questions to test your progress.
Check your answers at the back of the book on page 94.

1. Write down a formula for the total cost T in pence of y balloons at 85 pence each and 8 party poppers at z pence each.

   ....

2. Here are some patterns.

   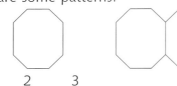 

   1   2   3

Pattern Number	1	2	3	4	5
Number of edges	8	15	22		

   a) Complete the table.
   b) How many edges will there be in the 15th pattern?
   c) Write down a formula connecting the pattern number (p) and the number of edges (e).

3. Solve the following equations:
   a) $2x + 4 = 10$ b) $3x - 1 = 11$ c) $5x - 3 = 2x + 12$ d) $3(x + 1) = 9$ e) $2(x + 1) = x + 3$

   ....

4. The perimeter of the triangle is 22 cm.
   a) Write down an equation for the perimeter of the triangle.

   ....

   b) Use your equation to find the length of the shortest side of the triangle.

   ....

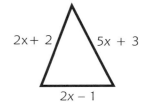

$2x + 2$ $5x + 3$

$2x - 1$

5.
   a) Complete the table of values for $y = 3x - 4$.

x	−2	−1	0	1	2	3
$y = 3x - 4$						

   b) On graph paper plot your values for x and y. Join your points with a straight line.
   c) Write down the coordinates of the points where your line crosses the x-axis.

   ....

6. Here are the first five terms of a sequence: 9, 13, 17, 21, 25.
   Write down an expression for the nth term of the sequence.

7. Multiply and simplify the following:
   a) $2(t - 4) + 7$
   b) $12 - 3(x - 2)$
   c) $4(3 + 2w) - 4w$

8. Factorise the following:
a) $16g - 40$
b) $12 - 4x$
c) $j^2 + 4j$
d) $28q^2 + 21q$

9. $v = 2p - 5q$
a) Find v when $p = 7$ and $q = 3$.
b) Find v when $p = -12$ and $q = -2$.

10. Amy is x years old. Amy's brother, Ben, is 3 years older than her; and Cara, her sister, is 2 years younger than her. The sum of their ages is 28. Find out x, Amy's age.

11. Solve the following inequalities
a) $7 \geq x + 3$
b) $2x - 4 \leq 24$

12. For each straight-line equation write down the gradient, m, and where it crosses the y-axis, c.
a) $y = 4x + 7$
b) $y = 3 - \frac{1}{4}x$
c) $5 - 2x = y$

13. Water is being poured into these containers at a rate of 250 ml per second. The graphs below show how the height of the water changes with time. Match the containers with the graphs.

A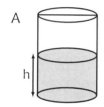

B

C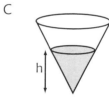

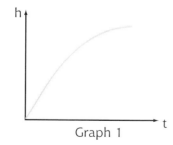
Graph 1

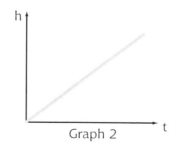
Graph 2

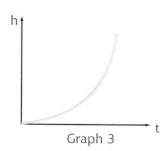
Graph 3

14. Draw the following straight lines
a) $y = 2x + 4$
b) $y = 7 - x$

How did you do?

1–5	correct	.... start again
6–9	correct	.... getting there
10–12	correct	.... good work
13–14	correct	.... excellent

Metric units

Metric units

Converting units

- If changing from **small** units to **large** units (for example, g to kg), **divide**.
- If changing from **large** units to **small** units (for example, km to m), **multiply**.

Length	Weight	Capacity
10 mm = 1 cm	1000 mg = 1 g	1000 ml = 1 l
100 cm = 1 m	1000 g = 1 kg	100 cl = 1 l
1000 m = 1 km	1000 kg = 1 tonne	1000 cm³ = 1 l

km　m　cm　mm
× 1000　× 100　× 10

Examples

a) Change 50 mm into cm.
mm are smaller than cm, so divide by the number of mm in a cm.
50 ÷ 10 = 5 cm.

b) Change 6 km into mm.
km are bigger than mm, so multiply by the number of m in a km, cm in a m and mm in a cm.
6 × 1000 × 100 × 10 = 6 000 000 mm.

Accuracy of measurement

There are two types of measurements: discrete measurements and continuous measurements.

Discrete measures

These are quantities that can be counted; for example, the number of baked bean tins on a shelf.

Continuous measures

These are measurements which have been made by using a measuring instrument; for example, the height of a person. Continuous measures are **not exact**.

Example

Nigel weighs 72 kg to the nearest kg. His actual weight could be anywhere between 71.5 kg and 72.5 kg.

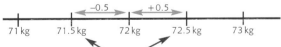

These two values are the **limits** of Nigel's weight.

If W represents weight, then $71.5 \leq W < 72.5$

Another way of representing $71.5 \leq W < 72.5$ is 72 ± 0.5

This is the lower limit of Nigel's weight (sometimes known as the lower bound). Anything below 71.5 would be recorded as 71 kg.

This is the upper limit (upper bound) of Nigel's weight. Anything from 72.5 upwards would be recorded as 73 kg.

Example

The length of a seedling is measured as 3.7 cm to the nearest tenth of a cm. What are the upper and lower limits of the length?

lower limit

upper limit

$3.65 \leq L < 3.75$ is 3.7 ± 0.05

2D shapes

Top Tip
See page 51 for explanations of symmetry.

2D shapes

Triangles

There are several types of triangle.

Equilateral
3 sides equal.
3 angles equal.

Isosceles
2 sides equal.
Base angles equal.

Right-angled
Has a 90° angle.

Scalene
No sides or angles the same.

Quadrilaterals

These are four-sided shapes.

Square
4 lines of symmetry.
Rotational symmetry of order 4.

Rectangle
2 lines of symmetry.
Rotational symmetry of order 2.

Parallelogram
No lines of symmetry.
Rotational symmetry of order 2.

Rhombus
2 lines of symmetry.
Rotational symmetry of order 2.

Kite
1 line of symmetry.
No rotational symmetry.

Trapezium
Isosceles trapezium:
1 line of symmetry.
No rotational symmetry.

Areas of quadrilaterals and triangles

Area of a rectangle

Area = length × width
$A = l \times w$

Area of a parallelogram

Area = base × perpendicular height
Remember to use the perpendicular height, not the slant height.
$A = b \times h$

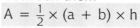

Area of a triangle

Area = $\frac{1}{2}$ base × perpendicular height

$A = \frac{1}{2} \times b \times h$

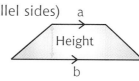

Area of a trapezium

Area = $\frac{1}{2}$ × (sum of parallel sides) × perpendicular height between them
$A = \frac{1}{2} \times (a + b) \times h$

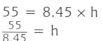

Examples

Find the areas of the following shapes, giving the answers to 3 s.f. where necessary.

a) A = b × h
= 12 × 4
= 48 cm²

b) A = $\frac{1}{2}$ × (a + b) × h
= $\frac{1}{2}$ × (4.9 + 10.1) × 6.2
= 46.5 cm²

Example

If the area of this triangle is 55 cm², find the height, giving your answer to 3 s.f.

A = $\frac{1}{2}$ × b × h
55 = $\frac{1}{2}$ × 16.9 × h
55 = 8.45 × h
$\frac{55}{8.45}$ = h

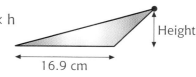

Substitute the values into the formula.
Divide both sides by 8.45.
So h = 6.51 cm (to 3 s.f.)

3D shapes

3D solids

Cube

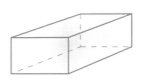
Cuboid

Square-based pyramid

Triangular-based pyramid (tetrahedron)

Cylinder

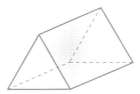
Triangular prism

Cone

Sphere

Volume of prisms

Volume of a cuboid
volume = length × width × height
V = L × w × h

> A prism is any solid which can be cut into slices, which are all the same shape. This is called having a uniform cross-section.

Volume of a prism
volume = area of cross-section × length
V = a × L

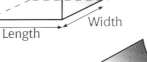
Area of cross section — Length

> To find the surface area of a cuboid work out the area of each face and then add them together. SA = 2lh + 2wh + 2lw.

Volume of a cylinder
Cylinders are prisms where the cross-section is a circle.
volume = area of cross-section × length

$$V = \pi r^2 \times h$$
area of circle   height or length

> The area of a cross-section of a triangular prism is the area of a triangle.

Radius
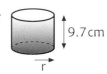
Height

Examples

Find the volumes of the following 3D shapes, giving the answer to 3 s.f. Use $\pi = 3.14$.

a) V = a × L
= $(\frac{1}{2} \times b \times h) \times L$
= $(\frac{1}{2} \times 9.6 \times 7) \times 15.1$
= 507.36 cm³
= 507 cm³ (3 s.f.)

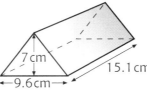
7 cm
9.6 cm
15.1 cm

b) V = $\pi r^2 \times h$
= $3.14 \times 10.7^2 \times 24.1$
= 8663.92 cm³
= 8660 cm³ (3 s.f.)

10.7 cm
24.1 cm

Example

If the volume of the cylinder is 500 cm³, work out the radius. (Use $\pi = 3.14$.)

9.7 cm
r

V = $\pi r^2 \times h$
$500 = 3.14 \times r^2 \times 9.7$ Substitute into the formula.

$500 = 30.458 \times r^2$
$\frac{500}{30.458} = r^2$ 30.458 Divide both sides by 30.458.

$r^2 = 16.416 ...$ Take the square root to find the radius.
$r = \sqrt{16.416} ...$ So r = 4.05 cm (3 s.f.)

Converting volume units

Another tricky topic which usually catches people out!

Example

The cube has a length of 1 m – this is the same as a length of 100 cm.

Therefore $1 m^3 = 1\,000\,000\,cm^3$

– not quite what you may have expected!

It's necessary therefore to change all the lengths to the same unit before starting a question!

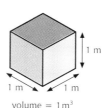

1 m
1 m   1 m
volume = $1 m^3$

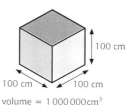

100 cm
100 cm   100 cm
volume = $1\,000\,000\,cm^3$

Dimensions

- The dimension of **perimeter** is **length** (L); it is a measurement in one dimension.
- The dimension of **area** is **length × length** ($L \times L = L^2$); it is a measurement in two dimensions.
- The dimension of **volume** is **length × length × length** ($L \times L \times L = L^3$); it is a measurement in three dimensions.
- Values like 3, π, 6.2, etc. have no dimensions.

> A formula with a mixed dimension is impossible.

Examples

The letters a, b, c and d all represent lengths. For each expression, write down whether it represents a length, area or volume.

a) $a^2 + b^2$ = (length × length) + (length × length) = area
$$L^2 + L^2$$

b) $\frac{1}{3} \pi abc$ = number × length × length × length = volume
$$L^3$$

> A dimension greater than 3 is impossible, so it has no dimension.

c) $2 \pi a + \frac{3}{4} \pi d$ = (number × length) + (number × length) = length
$$L + L$$

d) $\frac{5}{9} \pi a^2 d + \pi b^2 c^2$ = (number × length × length × length) + (number × length² × length²) = none
$$L^3 + L^4$$

Quick Test

1. Work out the volumes of the following 3D shapes. Give your answer to 3 s.f.

   a)

   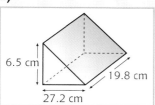

   6.5 cm
   19.8 cm
   27.2 cm

   b)

   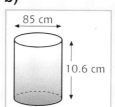

   85 cm
   10.6 cm

2. How many cm^3 are in $1.4 m^3$?

3. X, Y, Z represent lengths. For each expression write down whether it could represent perimeter, area or volume.

   a) $\sqrt{X^2 + Y^2 + Z^2}$

   b) $\frac{5}{9} \pi X^3 + 2Y^3$

   c) $\frac{1}{3} \frac{XYZ^3}{Y}$

Nets

Nets of solids

If you were making the shape you would need tabs for sticking.

The net of a 3D shape is the 2D shape which is folded to make the 3D shape.

Examples

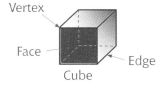

Vertex

Face

Cube

Edge

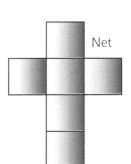

Net

Square-based pyramid

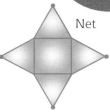

Net

Surface area of cuboids

To find the surface area of a cuboid, the dimensions must be known.

Height

Breadth

Length

Surface Area = 2((Length × Breadth) + (Length × Height) + (Breadth × Height))

In a cuboid, opposite sides are identical. That is why we find the sum of the areas of the three different rectangles and then double the sum.

Example

Find the surface area of the cuboid

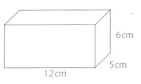

6 cm

5 cm

12 cm

Surface Area = 2((12 × 5) + (12 × 6) + (5 × 6))

= 2(60 + 72 + 30)

= 2(162)

= 324 cm²

Quick Test

1. Find the surface area of the cuboid

6 cm

5 cm

3 cm

Symmetry

Reflective symmetry

Both sides of a symmetrical shape are the same when a mirror line is drawn across it. The mirror line is known as the **line** or **axis of symmetry**.

1 line

1 line

3 lines

No lines

Rotational symmetry

A 2D (two-dimensional) shape has rotational symmetry if, when turned, it looks exactly the same. **The order of rotational symmetry** is the number of times the shape turns and looks the same. For the letter M the shape has 1 position. It is said to have **rotational symmetry of order 1**, or no rotational symmetry.

Order 1

Order 1

Order 3

Order 4

Plane symmetry

This is symmetry in 3D (three-dimensional) solids only.

A 3D shape has a **plane of symmetry** if the plane divides the shape into two halves, and one half is an exact mirror image of the other.

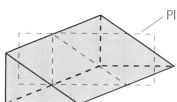
Plane of symmetry

Top Tip
When asked to draw in a plane of symmetry on a solid, make sure that it is a closed shape – don't just draw in a line of symmetry.

Quick Test

1. What are the names of the three types of symmetry?

2. The dashed lines are the lines of symmetry. Complete the shape so that it is symmetrical.

3. Draw a plane of symmetry on this solid.

Answers 1. Plane, rotational, reflective 2. 3.

Circles

The circle

Diameter = 2 × radius

The **circumference** is the distance around the outside edge.

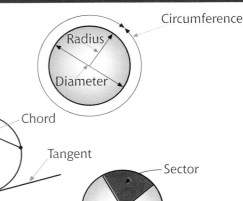

A **chord** is a line that joins two points on the circumference.
The line does not go through the centre.

A **tangent** touches the circle at one point only.

An **arc** is part of the circumference.

A **sector** of a circle is the area bounded by two radii and an arc.

A **segment** of a circle is the area bounded by a chord and an arc.

Circumference and area of a circle

circumference = π × diameter

$C = \pi \times d$
$= \pi \times \text{radius} \times 2$
$= \pi \times r \times 2$

area = π × (radius)²

$A = \pi \times r^2$

Example

The diameter of a circular rose garden is 5 m.

Find the circumference and area of the rose garden.

$C = \pi \times d$ Substitute in the formula.

$= 3.14 \times 5$ Use $\pi = 3.14$ or the value of π on your calculator.

$= 15.7\,\text{m}$ **EXP** often gives the value of π.

When finding the area, work out the radius first.

$d = 2 \times r$ so $r = d \div 2$, and $r = 2.5$ m

$A = \pi \times r^2$

$= 3.14 \times 2.5^2$ Remember 2.5^2 means 2.5×2.5.

$= 19.625$

$= 19.6\,\text{m}^2$ (3 s.f.)

Note – this answer could be left as 6.25π, i.e. in terms of π.

Example

A circle has an area of 40 cm². Find the radius of the circle, giving your answer to 3 s.f.
Use $\pi = 3.14$.

$A = \pi \times r^2$

$40 = 3.14 \times r^2$ Substitute the values into the formula.

$\frac{40}{3.14} = r^2$ Divide both sides by 3.14.

$r^2 = 12.738\,\ldots$

$r = \sqrt{12.738\,\ldots}$ Take the square root
$= 3.57$ cm (3 s.f.)   to find r.

Remember r^2 means $r \times r$

Circle theorems

There are several theorems you need to know and be able to apply. The theorems are:

1. The perpendicular bisector of any chord passes through the centre.

2. The angle in a semicircle is always 90°.

3. The radius and a tangent always meet at 90°.

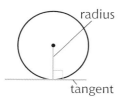

4. Angles in the same segment are equal,
e.g. A$\hat{B}$C = A$\hat{D}$C

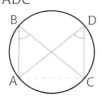

5. The angle at the centre is twice the angle at the circumference,
e.g. P$\hat{O}$Q = 2 × P$\hat{R}$Q

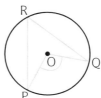

6. Opposite angles of a cyclic quadrilateral add up to 180°.
(A cyclic quadrilateral is a 4-sided shape with each corner touching the circle),
i.e. x + y = 180°
$\quad\quad$ a + b = 180°

7. The lengths of two tangents from a point are equal,
e.g. RS = RT

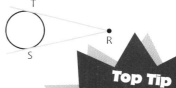

Top Tip
It is very important that you learn all of the seven theorems listed and that you can apply and explain them in the exam.

Quick Test

1. Work out the circumference and the area of each of the following circles

a)

8 m

b)
12.5 cm

2. Work out the area of the shaded region.

← 10 cm →

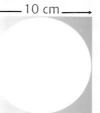

3. Calculate the missing angles in the diagram below.

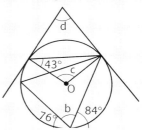

Angles

An acute angle is between 0° and 90°.	**An obtuse angle is between 90° and 180°.**	**A reflex angle is between 180° and 360°.**	**A right angle is 90°.**

Angle facts

Angles on a **straight line** add up to **180°**.
a + b + c = 180°

Angles at a **point** add up to **360°**.
a + b + c + d = 360°
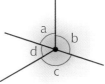

Angles in a **triangle** add up to **180°**.
a + b + c = 180°

Angles in a **quadrilateral** add up to **360°**.
a + b + c + d = 360°

Vertically opposite angles are **equal**.
a = b, c = d
a + d = b + c = 180°

An **exterior angle** of a triangle equals the sum of the **two** opposite **interior angles**.
a + b = c

Reading angles

When asked to find XYZ or ∠XYZ or XŶZ, find the **middle letter** angle, angle Y.

Angle in parallel lines

 Alternate (z) angles are equal.

Corresponding angles are equal.

 Supplementary angles add up to 180°. c + d = 180°

Examples

Find the angles labelled by letters.

a = 50° + 70°
a = 120°

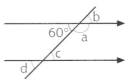

a + 80° + 40° + 85° = 360°
a = 360° – 205°
a = 155°

a = 120° (angles on a straight line)
b = 60° (vertically opposite)
c = 60° (corresponding to b)
d = 60° (vertically opposite to c)

Polygons

These are 2D shapes with **straight** sides.

Regular polygons are shapes with all sides and angles equal.

Number of sides	Name of polygon
3	Triangle
4	Quadrilateral
5	Pentagon
6	Hexagon
7	Heptagon
8	Octagon
9	Nonagon
10	Decagon

Tessellations

A **tessellation** is a pattern of 2D shapes which fit together without leaving any gaps.

For shapes to tessellate, the angles at each point must add up to 360°.

Example

Regular pentagons will not tessellate.
Each interior angle is 108°, and
3 × 108° = 324°.
A gap of 360° − 324° = 36° is left.

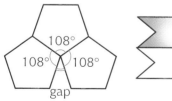

Angles in a polygon

There are two types of angle in a polygon:
interior (inside) and **exterior** (outside).

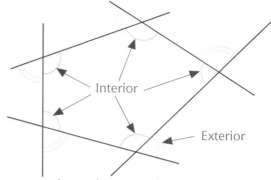

For a regular polygon with n sides:
- sum of exterior angles = 360°,
  so exterior angle = $\frac{360°}{n}$
- interior angle + exterior angle = 180°
- sum of interior angles = **(n − 2) × 180°**
  or **(2n−4) × 90°**

Example

A regular polygon has an interior angle of 150°.
How many sides does it have?
Let n be the number of sides.
exterior + interior = 180°
exterior angle = 180° − 150° = 30°
But exterior angle = $\frac{360°}{n}$
So n = $\frac{360°}{\text{exterior angle}}$
n = $\frac{360°}{30°}$ = 12

Top Tip
Make sure that you show full working out when carrying out an angle calculation. If you are asked to 'Explain', always refer to the angle properties, e.g. angles on a straight line add up to 180°, rather than how you worked it out.

Quick Test

1. Find the sizes of the angles below labelled by letters.

a)

b)

c)

2. Find the size of an
   a) exterior and
   b) interior angle of a regular pentagon.

Bearings and scale drawings

Bearings

- A **bearing** is the direction travelled between two points, given as an angle in degrees.
- All bearings are **measured clockwise** from the north line.
- All bearings should be given as 3 figures, e.g. 225°, 043°, 006°.

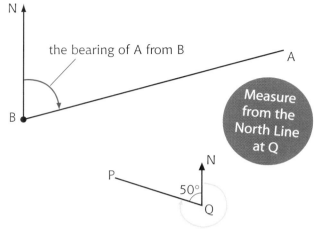

the bearing of A from B

Measure from the North Line at Q

Examples

Bearing of P **from** Q
= 060°

Bearing of P **from** Q
= 180° − 30° = 150°

Bearing of P **from** Q
= 360° − 50° = 310°

Back bearings

When finding the **back bearing** (the bearing of Q **from** P above):

- draw in a north line at P;
- the two north lines are parallel lines, so the angle properties of parallel lines can be used.

Measure from the north line at P

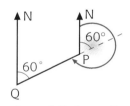

Bearing of Q **from** P
= 60° + 180° = 240°

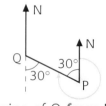

Bearing of Q **from** P
= 360° − 30° = 330°

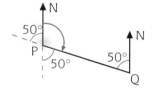

Bearing of Q **from** P
= 180° − 50° = 130°

Top Tip
The word **from** is important when answering bearing questions. It tells you where to put the north line and where to measure.

Scale drawings and bearings

Scale drawings are useful for finding lengths and angles.

Example

A ship sails from a harbour for 15 km on a bearing of 040°, and then continues due east for 20 km. Make a scale drawing of this journey using a scale of 1 cm to 5 km. How far will the ship have to sail to get back to the harbour by the shortest route?

What will the bearing be?

Shortest route = 6.4 × 5 km = 32 km
Bearing = 70° + 180° = 250°

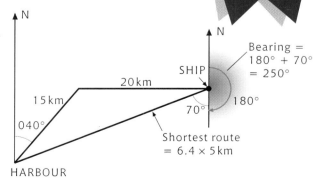

Bearing =
180° + 70°
= 250°

SHIP

20 km

15 km

040°

70°

180°

Shortest route
= 6.4 × 5 km

HARBOUR

Note – this diagram is not drawn accurately but is used to show you what your diagram should look like

Scales and maps

Scales are often used on maps.
They are usually written as a ratio.

Example

The scale on a road map is 1 : 250 000.
Edinburgh and Glasgow are 28 cm apart on the map.
Work out the real distance, in km, between Edinburgh and Glasgow.

Scale 1 : 250 000, distance on map is 28 cm.
Real distance = 28 × 250 000 = 7 000 000 cm

Divide by 100 to change cm to m.
7 000 000 ÷ 100 = 70 000 m

Divide by 1000 to change m to km.
70 000 ÷ 1000 = 70 km

> A scale of 1 : 250 000 means that 1 cm on the scale drawing represents a real length of 250 000 cm.

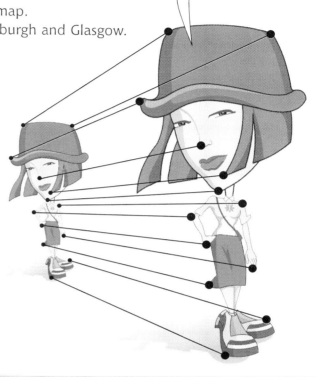

Quick Test

1. What are the bearings of A from B in the following:

a)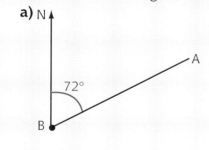

b)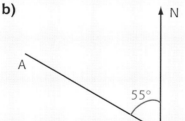

c)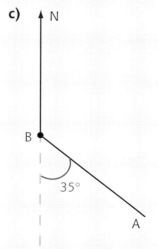

2. For each of the questions above work out the bearing of B from A.

3. The scale on a road map is 1 : 50000. If two towns are 14 cm apart on the map, work out the real distance between them.

Similarity

Similar shapes

Similar figures are those which are the **same shape** but **different sizes**.

Corresponding angles are equal.
Corresponding lengths are in the same ratio.

Corresponding
angles are equal.

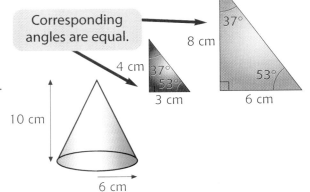

Examples

Notice corresponding lengths are in the same ratio.
The lengths of the bigger cone are
twice the size of the smaller cone.

Working out the scale factor

- When a shape gets bigger the scale factor is an **enlargement**.
- An enlargement scale factor is greater than 1.
- When a shape gets smaller the scale factor is a **reduction**.
- A reduction scale factor is less than 1.

Example

To go from shape A to B, the scale factor will be an enlargement.

Enlargement Scale Factor $= \frac{\text{Length of B}}{\text{Length of A}} = \frac{\text{Breadth of B}}{\text{Breadth of A}} = \frac{9}{6} = \frac{15}{10} = 1.5$

To go from shape B to A, the scale factor will be a reduction.

Reduction Scale Factor $= \frac{\text{Length of A}}{\text{Length of B}} = \frac{\text{Breadth of A}}{\text{Breadth of B}} = \frac{6}{9} = \frac{10}{15} = \frac{2}{3}$

Top Tip
Always make sure you put the corresponding sides in the correct order and remember that whatever you are trying to work out must go on the top of the fraction otherwise you will have a tricky calculation to do.

Finding missing lengths of similar figures

It is useful to draw the
two triangles first.

Example

Find the missing length, a, giving your answer to 2 s.f.

$\frac{a}{11} = \frac{9}{14}$ Corresponding sides are in the same ratio.

$a = \frac{9}{14} \times 11$ Multiply both sides by 11.

$a = 7.1\,\text{cm}$ (2 s.f.)

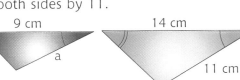

Example

Calculate the missing length.

$\frac{x}{6} = \frac{24}{16}$ Corresponding sides are in the same ratio.

$x = \frac{24}{16} \times 6$ Multiply both sides by 6.

$x = 9\,\text{cm}$

Measurements in cm

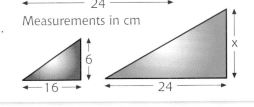

Sample exam-style questions

1. John's model boat is made to a scale of 1:50.

a. The model is 14 cm long. Calculate the length of the boat in metres.
real size of boat = 50 × 14 = 700 cm   700 ÷ 100 = 7 m

b. The actual boat's height is 2.5 m. Calculate the height of the model in cm.
height of model boat = 2.5 ÷ 50 = 0.05 m   0.05 × 100 = 5 cm

2. Find the length of x

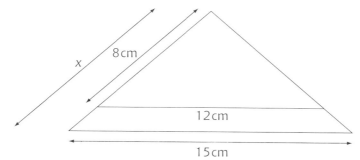

Draw the separate triangles

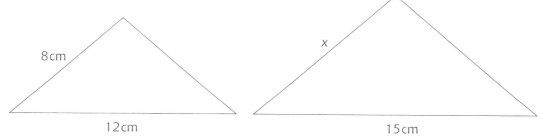

x is part of the large triangle so the scale factor will be an enlargement

$$\frac{x}{8} = \frac{15}{12}$$

$$x = \frac{15}{12} \times 8 = 1.25 \times 8 = 10 \text{ cm}$$

Quick Test

1. Find the lengths labelled by the letters in these similar shapes.

a)

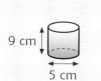

b)

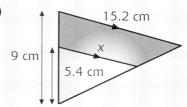

Pythagoras' theorem

The hypotenuse is the longest side of a right-angled triangle. It is always opposite the right angle.

Pythagoras' theorem states: in any right-angled triangle, the square on the hypotenuse is equal to the sum of the squares on the other two sides.

Using the letters in the diagram, the theorem is written as:

$$c^2 = a^2 + b^2$$

This may be rearranged to give $a^2 = c^2 - b^2$ or $b^2 = c^2 - a^2$, which are useful when calculating one of the shorter sides.

Finding the length of the hypotenuse

Remember: Pythagoras' theorem can only be used for right-angled triangles.

Step 1 Square the two numbers that you are given.

Step 2 To find the hypotenuse (longest side) add these two squared numbers.

Step 3 After adding the two lengths find the square root of your answer.

Example

Find the length of AB, giving your answer to 1 decimal place.

Using Pythagoras' theorem gives:

$$
\begin{aligned}
AB^2 &= AC^2 + BC^2 \\
&= 12^2 + 14.5^2 \\
&= 354.25 \\
AB &= \sqrt{354.25} \qquad \text{Square root to find AB.} \\
&= 18.8\,\text{m (to 1 d.p.)} \quad \text{Round to 1 d.p.}
\end{aligned}
$$

Finding the length of a shorter side

Follow the steps for finding the hypotenuse except:

Step 2 To find the shorter length subtract the smaller value from the larger value.

- Remember to square root ($\sqrt{}$) your answer.

Example

Find the length of FG, leaving your answer in surd form.

Using Pythagoras' theorem gives:

$$
\begin{aligned}
EF^2 &= EG^2 + FG^2 \\
FG^2 &= EF^2 - EG^2 \\
&= 92 - 82 \\
&= 17 \\
FG &= \sqrt{17}
\end{aligned}
$$

Rearrange the formula:
use $a^2 = c^2 - b^2$

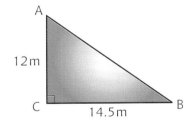

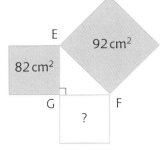

Leaving an answer in surd form just means to leave it with a square root.

Top Tip

Pythagoras' theorem allows us to calculate the length of one of the sides of a right-angled triangle, when the other two sides are known. If you are not told to what degree of accuracy to round your answer be guided by significant figures given in the question.

Calculating the length of a line *ab*, given two sets of coordinates

By drawing in a triangle between the two points A (1, 2) and B (7, 6) we can find the length of AB by Pythagoras' Theorem.

Horizontal distance = (7 − 1) = 6
Vertical distance = (6 − 2) = 4
Length of $(AB)^2 = 6^2 + 4^2$
$(AB)^2 = 36 + 16$
$(AB)^2 = 52$
$AB = \sqrt{52}$
Length of AB = 7.21

The **midpoint** of AB, M, has coordinates (4, 4)
i.e. $\frac{(1 + 7)}{2}, \frac{(2 + 6)}{2}$

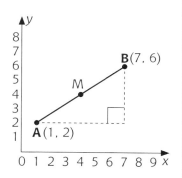

Solving problems

Example

- Calculate the height of this isosceles triangle.

Example

- A ladder of length 13 m rests against a wall. The height up the wall the ladder reaches is 12 m. How far away from the wall is the foot of the ladder?

$13^2 = x^2 + 12^2$
$13^2 - 12^2 = x^2$
$169 - 144 = x^2$
$x^2 = 25$
$x = \sqrt{25} = 5$ m

The foot of the ladder is 5 m away from the wall.

- Split the triangle down the middle to make it right-angled.

Using Pythagoras' theorem gives:
$5^2 = h^2 + 1.75^2$
$h^2 = 5^2 - 1.75^2$
$h^2 = 21.9375$
$h = \sqrt{21.9375} = 4.68$ cm (to 2 d.p.)

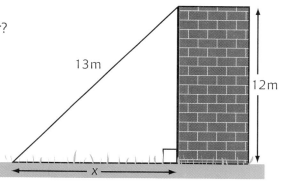

Quick Test

1. Calculate the lengths of the sides marked with a letter. Give your answers to 1 d.p.
   a)   b)

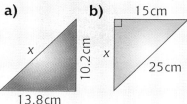

2. Work out the length of the diagonal on this rectangle.

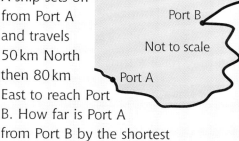

3. A ship sets off from Port A and travels 50 km North then 80 km East to reach Port B. How far is Port A from Port B by the shortest

Trigonometry in right-angled triangles

Trigonometry in right-angled triangles can be used to calculate an unknown angle or an unknown length.

Labelling the sides of the triangle

hyp (hypotenuse) is opposite the right angle.
opp (opposite side) is opposite the angle θ.
adj (adjacent side) is next to the angle θ.

θ is a Greek letter called **theta** and is used to represent **angles**.

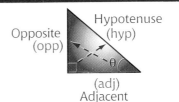

Trigonometric ratios

The three trigonometric ratios are:

$$\text{Sin } \theta = \frac{\text{opposite}}{\text{hypotenuse}} \qquad \text{Cos } \theta = \frac{\text{adjacent}}{\text{hypotenuse}} \qquad \text{Tan } \theta = \frac{\text{opposite}}{\text{adjacent}}$$

The made-up word **SOH CAH TOA** is a quick way of remembering the ratios. The word comes from the first letters of **S**in equals **O**pposite divided by **H**ypotenuse, etc.

To enter sin 30 into the calculator you usually do it backwards, i.e. [30] [SIN] (although some calculators do it forwards: check yours!)

Calculating the size of an angle

Example

Calculate angle AB̂C.

$\tan \theta = \frac{\text{opp}}{\text{adj}}$ Label the sides and decide on the ratio.

$\tan \theta = \frac{15}{27}$ Divide the top value by the bottom value.

$\tan \theta = 0.\dot{5}$

$\theta = 29.1°$ (1 d.p.)

On the calculator type in

[15] [÷] [27] [=] [INV] [TAN] [=]

You may have a shift key on your calculator.

> To find the angle you usually use the second function on your calculator.

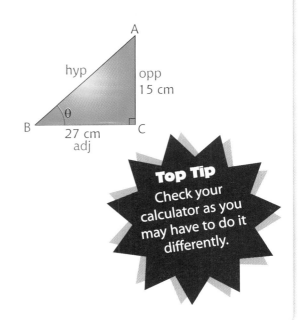

Top Tip
Check your calculator as you may have to do it differently.

Calculating the length of a side

Example

Calculate the length of BC.

- Label the sides first.
- Decide on the ratio.
  $\sin 30° = \frac{opp}{hyp}$
- Substitute in the values.
  $\sin 30° = \frac{BC}{25}$
  $25 \times \sin 30° = BC$ Multiply both sides by 25.
  $BC = 12.5\,cm$

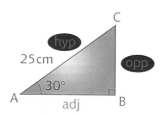

Top Tip
When calculating the size of an angle it should usually be rounded to 1 d.p. However do not round off your answer until right at the end of the question.

Example

Calculate the length of FG

$\tan 40 = \frac{opp}{adj} = \frac{GF}{20}$

$20 \times \tan 40 = GF$

$GF = 16.8\,cm$ (1 d.p.)

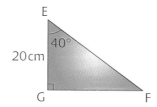

Quick Test

1. Label the sides of these triangles with respect to θ.

a) 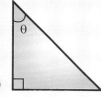   b)   c)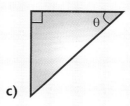

2. Calculate the length of x in each triangle.

a) 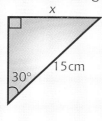   b)   c)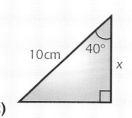

3. Work out the size of the angle in each of these triangles.

a) 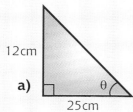   b)   c)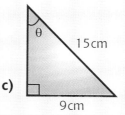

 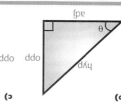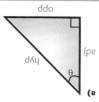

Applications of trigonometry

Angles of elevation and depression

The **angle of elevation** is measured from the horizontal **upwards**.

Angle of elevation

The **angle of depression** is measured from the horizontal **downwards**.

Angle of depression

Sample exam-style question

The diagram shows the position of four towns.
Beetown is due east of Diveton.
Crawley is 27 km from Beetown.
Ashcroft is 12 km due north of Diveton.
Beetown is on a bearing of 140° from Ashcroft.

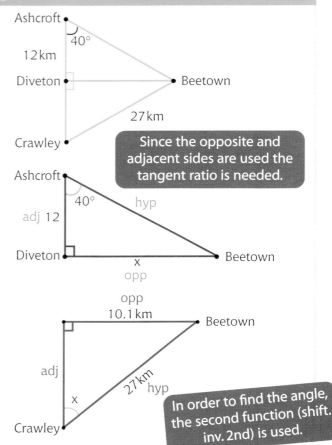

Since the opposite and adjacent sides are used the tangent ratio is needed.

a) Calculate the distance from Diveton to Beetown.
Give your answer in kilometres correct to 1 d.p.
$\tan 40° = \frac{opp}{adj}$ ✓
$\tan 40° = \frac{x}{12}$
$12 \times \tan 40° = x$ ✓
$x = 10.1$ km
10.1 km ✓

b) Calculate the bearing of Beetown from Crawley, giving your answer to the nearest degree.
$\sin x = \frac{opp}{hyp}$
$\sin x = \frac{10.1}{27}$
$\sin x = 0.374074$
$x = 21.97$ ✓
Bearing = 022°

Do not round off until the end of the question.

In order to find the angle, the second function (shift. inv. 2nd) is used.

Quick Test

Dipak stands 30 m from the base of a tower.
He measures the angle of elevation from ground level to the top of the tower as 50°. Calculate the height of the tower. Give your answer to 3 s.f.

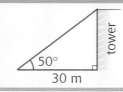

50°
30 m
tower

Answer 35.8 m (3 s.f.)

Sample exam-style questions

1.

a. Find the area of the triangle.

Area of a Triangle $= \frac{1}{2} \times b \times h$
$\frac{1}{2} \times 4 \times 2.5$
$2 \times 2.5 = 5\,m^2$

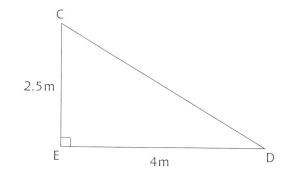

b. Find the value of angle ∠ECD.

We have to use trigonometry to find the angle
SOHCAHTOA.
We know the opposite length and the adjacent
length so we use the tan ratio.

$\tan \theta = \frac{opposite}{adjacent} = \frac{4}{2.5} = 1.6$
$\theta = 58.0°$ (to 1 d.p.)

2.

a. What is angle ∠ABC?

A triangle drawn in a circle with one of its
lengths being the same as the diameter of
the circle means it is a right-angled triangle.
∠ABC = 90°

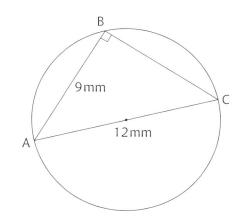

b. Find length BC.

Use Pythagoras to find length BC.
$a^2 + b^2 = c^2$
$9^2 + b^2 = 12^2$
$b^2 = 144 - 81$
$b^2 = 63$
$b = \sqrt{63} = 7.94\,mm$

c. Find angle ∠BAC.

Trigonometry has to be used to find the angle.
Since all three lengths are known in the triangle
you can choose from the three ratio's.
The cos ratio will be used here

$\cos \theta = \frac{adjacent}{hypotenuse} = \frac{9}{12} = 0.75$
$\theta = 41.4°$ (1 d.p.)

Test your progress

Use the questions to test your progress.
Check your answers at the back of the book on page 94.

1.

a) Draw the lines of symmetry on the rectangle.
b) What is the order of rotational symmetry of the rectangle?

....

2. Calculate the sizes of the angles marked with letters.

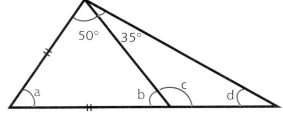

a) ....
b) ....
c) ....
d) ....

3. Write down the readings on these scales.

a)

b)
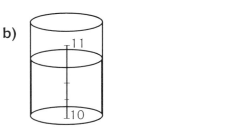

4.
a) Change 3.65 km to metres ....
b) Change 307 cm into metres. ....
c) Change 2.74 litres into millilitres. ....

5. The scale on a road map is 1 : 25 000. Amersham and Watford are 30 cm apart on the map. Work out the real distance in km between Amersham and Watford.

....

6. Wu Jing's weight is 60 kg to the nearest kilogram. Her actual weight could be anywhere between 59.5 kg and 60.5 kg. Write down the lower and upper limits for her weight of 60 kg.

7. Work out the areas of these shapes. Give your answer to 1 decimal place.

a)
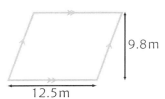

b)

....

8. A car travels a distance of 320 miles at an average speed of 65 mph. How long does it take?

....

9. Find angle x.
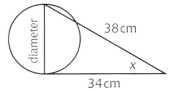

10. Jerry said that, 'The distance between Manchester and London is 240 miles to the nearest whole number.' Write down the smallest possible distance between Manchester and London.

....

11. A ladder of length 6 m rests so that the foot of the ladder is 3 m away from a wall. Calculate how far up the wall the ladder reaches. Give your answer to 2 significant figures.

....

....

12. Calculate the volume of the oil drum, clearly stating your units. Give your answer to three significant figures.

....

....

....

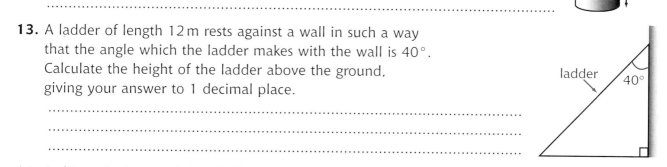

125 cm

Express Oil

1.58 m

13. A ladder of length 12 m rests against a wall in such a way that the angle which the ladder makes with the wall is 40°. Calculate the height of the ladder above the ground, giving your answer to 1 decimal place.

ladder 40°

....

....

....

14. A ship sails due north for 200 km. It then sails due west for 140 km. Calculate the bearing, to the nearest degree, of the ship from its starting point.

....

....

....

15. In the diagram MN is parallel to YZ, YMX and ZNX are straight lines, XM = 5.1 cm, XY = 9.5 cm, XN = 6.3 cm, YZ = 6.8 cm. ∠YXZ = 29°, ∠XZY = 68°

a) (i) Calculate the size of angle XMN ....

....

(ii) Explain how you obtained your answer ....

....

b) Calculate the length of MN ....

....

c) Calculate the length of XZ ....

....

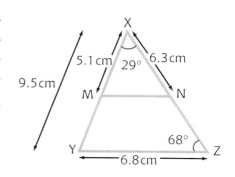

X

5.1 cm 29° 6.3 cm

9.5 cm

M   N

68°

Y   Z

6.8 cm

16. Here are some expressions:

$7r^2t$	$l\sqrt{r^2 + t^2}$	$\frac{rtl}{4}$	πr^2	$2\sqrt{r^2 + t^2}$	$2tl$	$4\frac{r^3}{t^2}$

The letters, r, t and l represent lengths. π, 2, 4 and 7 are numbers that have no dimensions.
Three of the expressions represent surface area. Tick the boxes (✔) underneath these three expressions.

How did you do?

1–5	correct	....	start again
6–10	correct	....	getting there
11–13	correct	....	good work
14–16	correct	....	excellent

Representing data

Stem and leaf diagrams

Stem and leaf diagrams are a way of recording information.

Example In a maths test, the following marks are recorded

78 55 65 74 99 42 94 90

90 65 60 79 56 53 45 97

When the information is put into a
stem and leaf diagram it looks like this ————————————➤

The figures on the left of the line form the stem. Each
figure on the right is called a leaf. The leaves increase
in value outwards from the stem. A key is needed at
the bottom. n, indicates the size of the data set and an
example of how to read the figures.

Mathematics Test Marks

```
4 | 2 5
5 | 3 5 6
6 | 0 5 5
7 | 4 8 9
8 |
9 | 0 0 4 7 9
```

$n = 16$ 5 | 6 represents 56%

Line graphs

These are a set of points joined by lines.

Year	1988	1989	1990	1991	1992	1993
Number of cars sold	2500	2900	2100	1900	1600	800

Middle values, like point A, have no meaning. A does not mean
that halfway between 1990 and 1991, there were 2000 cars sold.

> This example is known as a time series because
> the data is recorded at intervals of time

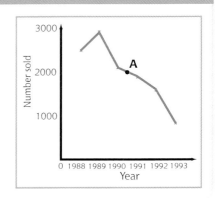

Drawing pie charts

These are used to illustrate data. They are circles split up into sections,
each section representing a certain number of items.

Example

The favourite sports of
24 students in year 11:

Sport	Frequency	Angle	Workings
Football	9	135°	$\frac{9}{24}$ x 360°
Swimming	5	75°	$\frac{5}{24}$ x 360°
Netball	3	45°	$\frac{3}{24}$ x 360°
Hockey	7	105°	$\frac{7}{24}$ x 360°
Total	24	360°	

To calculate the angles
for the pie chart:

- Find the total for the items listed.
- Find the fraction of the total for each item.
- Multiply the fraction by 360° to find the angle.

Key in on the calculator:
9 ÷ 24 × 3 6 0 =

Interpreting pie charts

Example

The pie chart shows some students travel to school. There are 18 students in total. How many travel by:

a) Car? b) Bus? c) Foot?

$$Car = \frac{60°}{360°} \times 18 = 3 \text{ students}$$

$$Bus = \frac{80°}{360°} \times 18 = 4 \text{ students}$$

$$Foot = \frac{220°}{360°} \times 18 = 11 \text{ students}$$

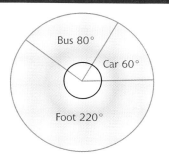

Bus 80°
Car 60°
Foot 220°

Histograms

Histograms are drawn to illustrate **continuous data**. They are similar to bar charts except that there are no gaps between the bars. The data must be grouped into equal class intervals if the length of the bar is used to represent the frequency.

Weight (kg)	Frequency
$45 \leq W < 55$	7
$55 \leq W < 65$	13
$65 \leq W < 75$	6
$75 \leq W < 85$	4

Example

The weights of 30 workers in a factory are shown in the table:

$45 \leq W < 55$, etc. are called **class intervals** – notice they are all equal in width. This means the weights are between 45 and 55 kg. A weight of 55 kg would be in the next group.

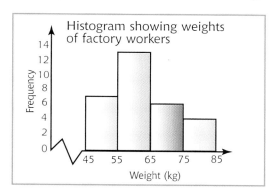

Note:

- The axes do not need to start at zero.
- The axes are labelled.
- The graph has a title.

Top Tip
When constructing a pie chart ensure that the calculated angles always add up to 360.

Quick Test

1. For this set of data, draw a pie chart.

Hair colour	Frequency
Brown	8
Auburn	4
Blonde	6
Black	6

Height (cm)	Frequency
$140 \leq h < 145$	
$145 \leq h < 150$	10
$150 \leq h < 155$	
$155 \leq h < 160$	
$160 \leq h < 165$	

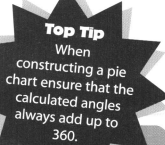

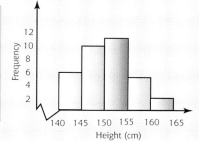

2. Using the histogram:

a) complete the frequency table.

b) How many people were in the survey?

Scatter diagrams and correlation

- A scatter diagram (scatter graph or scatter plot) is used to show two sets of data at the same time.
- Its importance is to show the correlation (connection) between two sets of data.

Types of correlation

There are three types of correlation

POSITIVE   **NEGATIVE**   **ZERO**

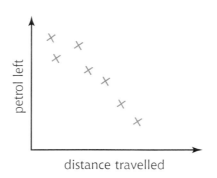 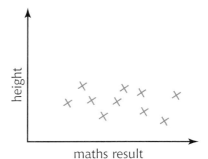

Positive

This is when both variables are increasing. If the points are nearly in a straight line there is said to be high positive correlation.

Negative

This is when one variable increases as the other decreases. If the points are nearly in a straight line there is said to be high negative correlation.

Zero

This is when there is little or no correlation between the variables.

Drawing a scatter diagram

Maths test (%)	64	79	38	42	49	75	83	82	66	61	54
History test (%)	70	36	84	70	74	42	29	33	50	56	64

The table shows the Maths and History test results of 11 pupils.

- Work out the scales first.
- Plot the points carefully
- Each time a point is plotted, tick it off.

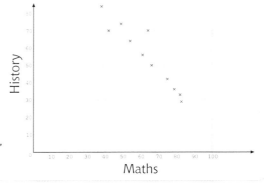

The scatter diagram shows that there is a strong negative correlation – in general, the better the pupils did in Maths, the worse they did in History and vice versa.

Lines of best fit

- **This is the line that best fits the data**. It goes in the direction of the data and has roughly the same number of points above the line as below it.
- A line of best fit can be used to make predictions.

Example

If Hassam was away for a Maths test but got 78% in History then from the scatter diagram we can estimate he would have got approximately 44% in Maths

⇨ Go to 78% on the History scale. Read across to the line then down.

Top Tip
When drawing the line of best fit, make sure you have about the same number of points above the line as below the line.

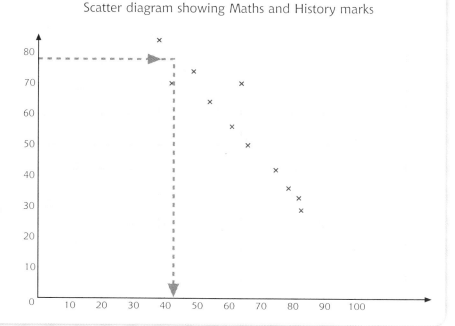

Scatter diagram showing Maths and History marks

Quick Test

For each pair of variables write down what type of correlation there is:

a) Number of pages in a magazine and the number of advertisements

b) The heights of students in a year group and the marks in the Maths test

c) The height up a mountain and the temperature

d) The age of a used car and its value

Answers 1. a) Positive **b)** Zero **c)** Negative **d)** Negative

Averages

'Average height of students is 163 cm'

Averages of discrete data

There are three types of average: mean, median and mode.

Mean: sometimes known as the 'average'. Symbol for the mean is $\bar{x}$.

$$\text{Mean} = \frac{\text{sum of a set of values}}{\text{the number of values used}}$$

Median: the middle value when the values are put in order of size.

Mode: the one that occurs the most often.

Range: this tells us how much the information is spread.
It is the highest value – lowest value.

Do not round off

Example

A football team scored the following number of goals in their first 10 matches:

2, 4, 0, 1, 2, 2, 3, 6, 2, 4

Find the mean, median, mode and range of the number of goals scored.

$$\text{Mean} = \frac{2+4+0+1+2+2+3+6+2+4}{10} = \frac{26}{10} = 2.6 \text{ goals}$$

Median =   0, 1, 2, 2, 2, 2, 3, 4, 4, 6 Put in order of size

$\cancel{0}, \cancel{1}, \cancel{2}, \cancel{2}, (2, 2) \cancel{3}, \cancel{4}, \cancel{4}, \cancel{6}$ Cross off from the ends to find the middle.

$$\frac{2+2}{2} = 2 \text{ goals}$$

Mode = 2 goals because it occurs 4 times.

Range = 6 – 0 = 6

Finding a missing number

If you are given the mean of a set of discrete data you can use the information to calculate a missing value.

Example:

The mean of 15, 17, y, 28 and 19 is 16.
What is the value of y?

$$\text{mean} = \frac{15+17+y+28+19}{5}$$

$$16 = \frac{79+y}{5}$$

so $16 \times 5 = 79+y$

so $y = 80-79$

$y = 1$

This just becomes a simple equation to solve

Finding averages from a frequency table

A frequency table tells us how many are in a group.

Example

Number of sisters (x)	0	1	2	3	4
Frequency (f)	4	9	3	5	2

This means that 2 people had 4 sisters.

If there are two numbers in the middle, the median is halfway between them.

Mean $(\bar{x})$ $= \dfrac{\Sigma fx}{\Sigma f}$

Σ means the sum of

$= \dfrac{(4\times0)+(9\times1)+(3\times2)+(5\times3)+(2\times4)}{4+9+3+5+2+0}$

$= \dfrac{38}{23} = 1.7$ (to 1 d.p.)

Median   Since there are 23 people who have been surveyed the median will be the 12th person.

11 people   12   11 people

The 12th person has 1 sister so the median = 1

Mode   This is the one with the highest frequency, that is 1 sister.

Range   4 − 0 = 4

Top Tip
When finding the mean from a frequency table try to remember to divide by the sum of the frequencies and not by how many groups there are.

Using appropriate averages

- The **mean** is useful when a 'typical' value is wanted.
  Be careful not to use the mean if there are extreme values.
- The **median** is a useful average to use if there are extreme values.
- The **mode** is useful when the most common value is needed.

Quick Test

1. Find the mean, median and mode for this set of data:

   12   15   21   10   8   12

2. Find the mean, median and mode for this set of data:

   7   3   6   9   12   8   10   12

3. Charlotte got a mean percentage mark of 80% for four subjects. She knew her marks for English (78%), Physics (54%) and History (97%).

   a) Calculate Charlotte's mark for her fourth subject, Music.

   b) Calculate the range in her test marks.

Answers 1. Mean = 13 Median = 12 Mode = 12 **2.** Mean = 8 Median = 9 Mode = 12 **3.** Music mark = 91% Range = 43

Probability 1

Probability is the chance or likelihood that something will happen.
All probabilities lie between 0 and 1.

```
0                          0.5                                1
Definitely will    Unlikely to         Very likely      Definitely
not happen         happen              to happen        will happen
```

What is probability?

Exhaustive events account for **all possible outcomes**.

For example, the list HH, HT, TH, TT gives all possible outcomes when two coins are tossed simultaneously.

Mutually exclusive events are events that **cannot happen at the same time**.

For example if two students are chosen at random:

Event A: one student has brown hair

Event B: one student wears glasses

These are not mutually exclusive because brown-haired students can wear glasses

Probability of an event $= \dfrac{\text{number of ways an event can happen}}{\text{total number of outcomes}}$

P(event) is a shortened way of writing probability of an event.

Example

There are 6 blue, 4 yellow and 2 red beads in a bag.
John chooses a bead at random. What is the probability he chooses:

a) a red bead

b) a yellow bead

c) a blue, yellow or red bead

d) a white bead

a) $P(\text{red}) = \frac{2}{12}$ or $\frac{1}{6}$

b) $P(\text{yellow}) = \frac{4}{12}$ or $\frac{1}{3}$

c) $P(\text{blue, yellow or red}) = \frac{12}{12} = 1$

d) $P(\text{white}) = 0$

Probability of an event not happening

If two events are mutually exclusive, then
P(event will happen) = 1 – P(event will not happen)

or

P(event will not happen) = 1 – P(event will happen)

Example

The probability that someone gets flu next winter is 0.42.
What is the probability that they do not get flu next winter?

P(not get flu) = 1 – P(get flu) = 1 – 0.42 = 0.58

Expected number

Example

If a fair die is thrown 300 times, approximately how many fives are likely to be obtained?

$P(5) = \frac{1}{6} \times 300 = 50$ fives

We multiply 300 by $\frac{1}{6}$ since a 5 is expected $\frac{1}{6}$ of the time

Example

The probability of passing a driving test at the first attempt is 0.65. If there are 200 people taking their test for the first time, how many do you expect to pass the test?

$0.65 \times 200 = 130$ people

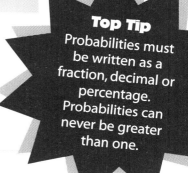

Relative frequencies

Relative frequencies can be used as an estimate of probability. If it is not possible to calculate probability, an experiment is used to find the relative frequency.

$$\text{Relative frequency of an event} = \frac{\text{number of times event occured}}{\text{total number of trials}}$$

Top Tip
Probabilities must be written as a fraction, decimal or percentage. Probabilities can never be greater than one.

Example

When a fair die was thrown 80 times a six came up 12 times.

What is the relative frequency of getting a six?

number of trials = 80   relative frequency $= \frac{12}{80} = 0.15$

number of sixes = 12

Quick Test

1. Write down an event that will have a probability of zero.

2. A box contains 3 salt 'n' vinegar, 4 cheese and 2 bacon flavoured packets of crisps. If a packet of crisps is chosen at random what is the probability that it is:

   a) salt 'n' vinegar? **b)** cheese? **c)** oinion flavoured?

3. The probability that it will not rain tomorrow is $\frac{2}{9}$.
   What is the probability that it will rain tomorrow?

4. The probability of achieving a grade C in Mathematics is 0.48.
   If 500 students sit the exam how many would you expect to achieve a grade C?

5. When a fair die was thrown 200 times, a five came up 47 times.
   What is the relative frequency of getting a five?

Probability 2

The multiplication law

When two events are **independent** the outcome of the second event is not affected by the outcome of the first. If two or more events are **independent**, the probability of A and B and C ... happening together is found by **multiplying** the separate probabilities.

$$P(A \text{ and } B \text{ and } C \ldots) = P(A) \times P(B) \times P(C) \ldots$$

Example

The probability that it will rain on any day in August is $\frac{3}{10}$. Find the probability that:

a) it will rain on both 1 August and 3 August

b) it will rain on 9 August but not 20 August.

a) $P(\text{rain and rain}) = \frac{3}{10} \times \frac{3}{10} = \frac{9}{100}$

b) $P(\text{rain and not rain}) = \frac{3}{10} \times \frac{7}{10} = \frac{21}{100}$

The addition law

If two or more events are **mutually exclusive** the probability of A or B or C ... happening is found by **adding** the probabilities.

$$P(A \text{ or } B \text{ or } C \ldots) = P(A) + P(B) + P(C) + \ldots$$

Example

There are 11 counters in a bag: 5 of the counters are red and 3 of them are white. Lucy picks a counter at random. Find the probability that Lucy's counter is either red or white.

$P(\text{red}) = \frac{5}{11}$ $\qquad P(\text{white}) = \frac{3}{11}$

$P(\text{red or white}) = P(\text{red}) + P(\text{white}) = \frac{5}{11} + \frac{3}{11} = \frac{8}{11}$ Red and white are mutually exclusive.

Tree diagrams

Tree diagrams are another way of showing the possible outcomes of two or more events. They may be written horizontally or vertically.

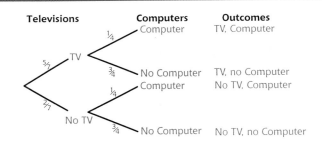

Example

In a class, the probability that a pupil will have his/her own television is $\frac{5}{7}$ and the probability that the pupil will have his/her own computer is $\frac{1}{4}$. These two events are independent. Draw a tree diagram of this information.

● Draw the first branch which shows the probabilities of having televisions.

● Put the probabilities on the branches.

● Draw the second branches which show the probabilities of having computers.

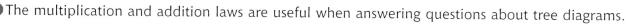

The multiplication and addition laws are useful when answering questions about tree diagrams.

a) Find the probability that a pupil will have their own TV and a computer.

P(TV and computer) = P(TV) x P(computer) = $\frac{5}{7} \times \frac{1}{4} = \frac{5}{28}$

b) Find the probability that a pupil will have only one of the items.

P(TV and no computer) = P(TV) x P(no computer) = $\frac{5}{7} \times \frac{3}{4} = \frac{15}{28}$

OR

P(no TV and computer) = P(no TV) x P(computer) = $\frac{2}{7} \times \frac{1}{4} = \frac{2}{28}$

P(only one of the items) = $\frac{15}{28} + \frac{2}{28} = \frac{17}{28}$

Sample space diagrams

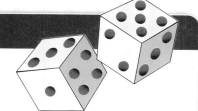

A table is helpful when considering outcomes of two events.
This kind of table is sometimes known as a **sample space diagram**.

Example

Two dice are thrown together and their scores are added. Draw a diagram to show all the outcomes. Find the probability of:

a) a score of 7

b) a score that is a multiple of 4

a) P(score of 7) = $\frac{6}{36} = \frac{1}{6}$

b) P(multiple of 4) = $\frac{9}{36} = \frac{1}{4}$

		First die				
	1	2	3	4	5	6
1	2	3	4	5	6	7
2	3	4	5	6	7	8
Second die 3	4	5	6	7	8	9
4	5	6	7	8	9	10
5	6	7	8	9	10	11
6	7	8	9	10	11	12

There are 36 outcomes

Quick Test

1. The probability the Meena does her homework is 0.8. The probability that Fiona does her homework is 0.45. Find the probability that both girls do their homework.

2. a) Draw a sample space diagram which shows the outcomes when two dice are thrown together and their scores are multiplied.

b) What is the probability of a score of 6?

c) What is the probability of a score of 37?

3. A bag contains 3 red and 4 blue counters. If a counter is taken out of the bag at random, its colour noted and then it is replaced, and a second counter is taken out, what is the probability of choosing a counter of either colour? (Hint: use a tree diagram to help you)

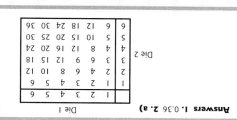

Answers 1. 0.36 2. a) [see diagram] b) $\frac{2}{36} = \frac{1}{16}$ c) 0 3. $\frac{24}{49}$

Test your progress

Use the questions to test your progress.
Check your answers at the back of the book on page 95.

1. The number of mm of rainfall that fell during the first eight days of August is shown below:

   a) Draw a line graph to display this information

   b) Work out the mean monthly rainfall for the first eight days of August.

Day	1	2	3	4	5	6	7	8
Rainfall	12	4	7	2	5	1	2	6

   ....

2. Reece carried out a survey to find out the favourite flavours of crisps of students in his class. The results are shown in the table (right)

   Draw a pie chart of this information.

Crisp flavour	Number of students
Cheese and onion	7
Salt and vinegar	10
Beef	6
Smokey bacon	1

3. Find the mean, median and mode of these quantities: **6, 2, 1, 4, 2, 2, 5, 3**

   ....

4. A bag contains 3 red, 4 blue and 6 green balls. If a ball is chosen at random from the bag, what is the probability of choosing:

   a) a red ball   b) a green ball   c) a yellow ball   d) a blue or red ball

   ....

5. The probability that Josie gets full marks on a table test is 0.82. What is the probability that she does not get full marks on the tables test?

   ....

6. A youth club has 75 members. 42 of the members are boys. There are 15 members who are boys under 13 years old. There are 21 members who are girls over 13 years old.

	Under 13 years old	13 years old and over	Totals
Boys			
Girls			
Total			

   a) Complete the two-way table.

   b) How many girls are under 13 years old?

   ....

7. The pie chart below shows how Erin spends a typical day.

   a) Measure the size of the angle for sleeping..

   b) Work out the number of hours that Erin works...

   c) For how many hours does Erin watch TV? ....

8. The probability of passing a driving test is 0.7. If 200 people take the driving test today, how many would you expect to pass?

   ....

9. Michelle is a swimmer. The probability of her winning a race is 84%. If she swims in 50 races this season, how many races would you expect her to win?

   ....

10. The number of people entering a supermarket is recorded hourly. The stem and leaf diagram shows the data collected.

    Customers

5	2	6		
6	0	2		
7				
8	4	5	9	9
9	1	1	2	5
10	8			

    a) List all the recordings in order of size ....

    b) For how many hours were records made? ....

    c) What is the mean number of customers? ....

    d) What is the median number of customers? ....

11. In a survey the heights of 10 girls and their shoe sizes were measured:

Height in cm	150	157	159	161	158	164	154	152	162	168
Shoe size	3	5	$5\frac{1}{2}$	6	5	$6\frac{1}{2}$	4	$3\frac{1}{2}$	6	7

    a) Draw a scatter diagram to illustrate this data.

    b) What type of correlation is there between height and shoe size? ....

    c) Draw a line of best fit on your diagram.

    d) From your scatter diagram, estimate the height of a girl whose shoe size is $4\frac{1}{2}$ ....

12. Two spinners are used in a game. The first spinner is labelled 2, 4, 6, 8. The second spinner is labelled 3, 5, 5, 7. Both spinners are spun. The score is found by multiplying the numbers on each spinner.

    a) Complete the table to show the possible scores:

    b) What is the probability of getting an even score? ....

    c) What is the probability of getting a score of 10? ....

		First spinner			
		2	4	6	8
Second spinner	3				
	5				
	5				
	7				

13. The weights of some high school students are displayed below:

    62kg 55kg 70kg 53kg 72kg 67kg 65kg 59kg 60kg 52kg

    a) Construct a stem and leaf diagram.

    b) What is the range in weights? ....

14. Ahmed and Matthew are going to take a swimming test. The probability that Ahmed will pass the swimming test is 0.85. The probability that Matthew will pass the swimming test is 0.6. The two events are independent.

    a) Complete the probability tree digram.

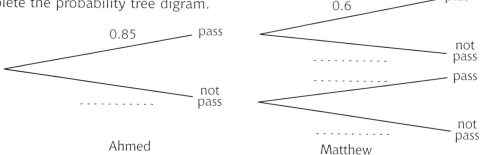

    Ahmed   Matthew

    b) Work out the probability that both Ahmed and Matthew will pass the swimming test ....

    c) Work out the probability that one of them will pass the swimming test and the other will not pass the swimming test ....

    How did you do?

1–5	correct	.... start again
6–12	correct	.... good work
13–14	correct	.... excellent

Whole numbers

Addition and subtraction

When performing addition or subtraction calculations you must ensure the numbers are lined up correctly.

Examples

1.
```
  4078
+ 5630
```

This has been written incorrectly because the bottom numbers are not in line with the top number.

Correct notation:
```
  4078
+ 5630
------
  9708
```

Multiplication

This is an essential skill. Most topics involve multiplication.

Examples

1.
```
  ²37
  ×4
-----
 148
```

When doing these calculations you multiply from right to left. In this example, the 4 multiplies the 7 to get 28. 8 is written down and 2 is then carried on. The 4 multiplies the 3 to give 12. Then the 2 is added on to the 12 to make 14. Then the 14 is written down.

2.
```
  ¹28
  ×12
-----
   56
  280
-----
  336
```

In this example the 2 multiplies the 8 to get 16. 6 is written down and 1 is then carried on. The 2 multiplies the 2 to give 4. Then the carried 1 is added on to the 4 to make 5 and this is written down. A zero is written in the units column of the second line, because the 1 means 1 ten and not 1 unit. The 1 multiplies the 8 to get 8, which is written down. The 1 multiplies the 2 to get 2, which is written down. The last step is to add the first and second rows together.

> **Top Tip**
> It is important that you know your times tables in order to answer non-calculator questions successfully.

Division

Division is the opposite of multiplication.

Examples

1. $72 \div 8 = 9$
2. $60 \div 12 = 5$

The above examples are quite straight forward. The difficulty is when there is a remainder in the answer.

3. $86 \div 4$

```
   21r2
 4│86
```

```
   21.5
 4│86.²0
```

This can be rewritten in the step notation. We cannot leave the answer as 21r2. When there is a remainder add in a decimal point and a zero after the point. Put the value of the remainder next to the zero, then do the division.

4. $724 \div 8$

```
    0 90.5
 8│7⁷24.⁴0
```

Quick Test

1. Find the following   a) 489 + 723   b) 367 − 329   c) 1287 − 749

2. Find the product of the following   a) 87 × 65   b) 625 × 73   c) 436 × 89

3. a) 69 ÷ 6   b) 124 ÷ 8   c) 148 ÷ 4

Answers 1. a) 1212 **b)** 38 **c)** 538 **2. a)** 5655 **b)** 45625 **c)** 38804 **3. a)** 11.5 **b)** 15.5 **c)** 37

Sample exam-style questions

Throughout the book there have been examples of how to perform decimal, fraction, percentages and negative problems without a calculator.

The questions below are common non-calculator questions.

1. $\frac{1}{3} + \frac{2}{9}$

First make the denominators the same

$\frac{3}{9} + \frac{2}{9} = \frac{5}{9}$

2. Find $\frac{2}{5}$ of 140

Remember 'of' means multiply

$\frac{2}{5} \times \frac{140}{1} = \frac{280}{5} = 56$

3. 12.4×0.8

Work out 124×8, ignoring the decimal points

```
  124
   ×8
 ────
  992
```

There is one number behind the decimal in 12.4 and one in 0.8, this means there will be 2 in the answer.
Move the decimal point 2 places $992 \rightarrow 9.92$ (answer)

4. Carry out the following calculation $6.8 - 3.85 + 4.79$

Do the subtraction first

```
  6.80
 −3.85
 ─────
  2.95
```

Now do the addition

```
  2.95
 +4.79
 ─────
  7.74
```

5. $0.296 \div 4 =$

$$4 \overline{\smash{\big)}\, 0.2^4 9^1 6} \quad = 0.0\,7\,4$$

Test your progress

Use the questions to test your progress.
Check your answers at the back of the book on page 94.

1. 5.25×40 ....

2. $15.25 + 12.8 - 13.7$ ....

3. 55% of 880 ....

4. 10.2×60 ....

5. David has a bag of fruit sweets. 2 are lemon, 4 are orange, 8 are apple and 6 are blackcurrent. What is the probability that he gets a blackcurrent sweet? (Simplify fully)

   ....

6. A quarter of Corinne's class are wearing a school uniform. If there are 32 pupils in the class, how many are not wearing a uniform?

   ....

7. Write the number out in full 2.7×10^{-3} ....

8. In the school basketball and football club, the ratio of basketball members to football members is 40:70

a) Express the ratio in its simplest form ....

b) The school sports department has been given £6600. This money will be divided between the basketball and football clubs in the same ratio as above. How much money will be allocated to the football club?

   ....

9. Find $\frac{3}{7}$ of 287 ....

10. Starting with the smallest, write the following in order. $\frac{2}{5}$ 41%   0.46 $\frac{1}{3}$ 23%   35%

    ....

11. $5 \times 3\frac{2}{5}$ ....

12. Find Angle x ....

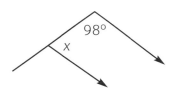

13. Gurmeet's pocket money is £40 per month. He spends one fifth of this on comic books, a quarter on the cinema and the rest he spends on clothes. How much does he spend on clothes?

    ....

14. At 8am the temperature was −7°C. At 3pm it was 12°C. What is the rise in temperature?

    ....

15. Write 4600 in standard form ....

16. 5% of £6.60 ....

17. 36×3.8 ....

18. $\frac{3}{10} \times \frac{2}{9}$ ....

The next pages have some example questions.
They are the sort of questions you'll come across in the exam.
We've done them for you so you can see how to solve them.

Number and money

The questions on the next 6 pages are typical of questions in the Standard Grade Maths exam at General Level. We have worked these out for you, so you can see how to solve questions like them in the exam.

Calculator questions have this icon

Non-calculator questions have this icon

Questions

Non-calculator

1. In a survey asking people's favourite juice drink, two-fifths of people who answered preferred apple juice. If 5400 people took part, how many of them preferred apple juice?

$\frac{2}{5}$ of 5400 $= \frac{2}{5} \times \frac{5400}{1} = \frac{10800}{5} = 2160$ people

Calculator

2. Elizabeth left her house at 08:30 and drove to work. She arrived at 09:20. If her average speed was 25mph, find the distance between her house and her work. Round your answer to one decimal place (1d.p.).

First find the time between 08:30 and 09:20   50 minutes

Then convert the time into hours as the speed is in mph (miles per hour). Do this by dividing the number of minutes taken (50) by the number of minutes in an hour (60).

$\frac{50}{60} = 0.83$ hours

Now we have the time and the speed we can work out the answer using the formula to find distance: Distance = Speed × Time

D = S × T

25 × 0.83 = 20.8 miles(1d.p.)

Non-calculator

3. a) The distance from Washington D.C. to Miami is 2,800 miles. Write this distance in standard form.

2.8×10^3

3. b) The distance from Paris to Washington D.C. is 6.09×10^3 miles. Write this distance as a normal number.

6090 miles

Calculator

4. After Easter, the price of chocolate eggs was reduced from £2 to £1.25. What was the percentage discount?

First work out the difference in price: £2 – £1.25 = £0.75

Percentage discount $= \frac{\text{discount value}}{\text{original value}} \times 100 = \frac{£0.75}{£2.00} \times 100 = 37.5\%$

Calculator

5. Max has a part-time job in a pet shop. His basic rate of pay is £6.80 per hour. At weekends he gets paid overtime at time-and-a-half. Last week he was paid £142.80, which included 6 hours overtime.

   How many hours did he work at the basic rate?

   First calculate his overtime pay by multiplying his hourly rate by 1.5 (time-and-a-half) and by 6 (hours).

   $6.80 \times 1.5 \times 6 = £61.20$

   Now work out how much he got paid for his basic wage by subtracting the amount of overtime from his total pay.

   $£142.80 - £61.20 = £81.60$

   Work out how many hours Max worked at the basic rate by dividing his basic wage by his hourly rate.

   $£81.60 \div £6.80 = 12$ hours

Calculator

6. a) A wide-screen TV is advertised for £380 + VAT. What is the total cost of the TV?

   VAT = 17.5%

   Find 17.5% of £380 = $0.175 \times £380$ = £66.50

   Add the VAT to the basic price for the total cost.

   £380 + £66.50 = £446.50

6. b) A week later the same wide-screen TV is put on sale for £350 including VAT. What is the percentage reduction on the TV (to 1 d.p.)?

   First work out the price reduction by subtracting the lower price from the higher price. Remember to compare both prices including VAT.

   £446.50 − £350 = £96.50

   Percentage discount = discount value/original value × 100
   = £96.50/£446.50 × = 21.6%

Non-calculator

7. 340 ÷ 0.8

   In this multiply everything by 10 in order to turn 0.8 into a whole number.

   $(340 \div 0.8) \times 10 = 3400 \div 8 =$
   $$8\overline{)3^34^20^40}\quad 0\;4\;2\;5$$

   So, 340 ÷ 0.8 = 425

Non-calculator

8. A 350-gram jar of summer fruits jam contains strawberries and raspberries. The ratio of strawberries to raspberries is 4 to 3. How many grams of raspberries will there be in the jar?

   First, work out what one part will be by dividing the total number of grams by the total number of parts.

   The ratio is 4:3 so there will be 7 parts in the whole.
   7 parts = 350g
   1 part = $\frac{350}{7}$ = 50g

   So, if raspberries are 3 parts, then the amount of raspberries will be
   3 × 50g = 150g

Algebra

Questions

Non-calculator

1. Solve

$$5x - 28 = 17$$
$$5x = 17 + 28$$
$$5x = 45$$
$$\frac{5x}{5} = \frac{45}{5}$$
$$x = 9$$

Non-Calculator

2. Write down the nth term for the following sequence   6, 10, 14, 18, 22

N = an + b
In the sequence 4 is being added on each time so a = 4.
N = 4n + b
When n = 1 N = 6 so use this to find out b.
6 = 4 + b
b = 2
Formula: N = 4n + 2

Non-Calculator

3. a) Complete the table below for
   $y = 2x - 2$

x	−3	0	3
y			

Substitute the values for x into the equation in order to find the three values for y.

$y = 2 \times -3 - 2 = -6 - 2 = -8$
$y = 2 \times 0 - 2 = 0 - 2 = -2$
$y = 2 \times 3 - 2 = 6 - 2 = 4$

x	−3	0	3
y	−8	−2	4

b) Using the table in **3.a)** draw the graph of the line $y = 2x - 2$.

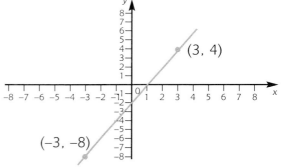

The line should cross the y-axis at −2.

The line should cross the x-axis at 1.

Non-Calculator

4. Factorise $x^2 - 4x$

x is a factor of each term, so we take this out as a common factor.

The expression is completed inside the bracket, so that the result is equivalent to $x^2 - 4x$ when multiplied out.

$x (x - 4)$

Calculator

5. a) An electrician charges a callout fee of £25 then £12 for each hour he works. Write a formula which describes the electrician's pay (P).

   This formula needs to contain the callout fee in it and the pay he receives for the hours he works.

   P = £25 + £12t (where t is the number of hours worked)

   b) How much will the electrician get paid for working 9 hours on a job?

   Substitute the number of hours worked for t to work out his pay.

   P = £25 + £12 × 9 = £25 + £108 = £133

Non-Calculator

6. Simplify $8y - 2(4 - 3y)$

   First multiply out the brackets $8y - 2(4 - 3y) = 8y - 8 + 6y$

   Then collect the like terms $8y + 6y - 8 = 14y - 8$

Non-Calculator

7. Solve $12x - 4 = 7x + 36$

   $12x - 7x = 36 + 4$

   $5x = 40$

   $x = 8$

Calculator

8. $j = 7\frac{h}{k}$ ☐   Find j when $h = 7$ and $k = 3.5$.

   $j = \frac{7 \times 7}{3.5} = \frac{49}{3.5} = 14$

Non-Calculator

9. Factorise $72 + 18y$

   9 is a factor of each term, so we take this out as a common factor.

   The expression is completed inside the bracket, so that the result is equivalent to $72 + 18y$ when multiplied out. So $9 (8 + 2y)$

Measure and shape

Questions

Calculator

1. Find the area of a circle with a diameter of 36 cm.
   Area of circle formula $= \Pi r^2$
   Diameter $= 36$ cm, so the radius will be half of this:
   radius $= 18$ cm
   Area $= \Pi \times 18^2 = 1017.9$ cm^2 (1 d.p.)

Calculator

1. Find the value for g

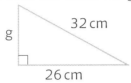

g

32 cm

26 cm

We need to use Pythagoras to find the missing side.

The side we want to find, g, is a short side.
$$26^2 + g^2 = 32^2$$
$$676 + g^2 = 1024$$
$$g^2 = 1024 - 676$$
$$g^2 = 348$$
$$g = 18.7 \text{ cm (1d.p.)}$$

Calculator

2. David wants to enlarge a photograph so that he can make it into a large poster. The length of the photograph is 8 cm and the breadth is 4.5 cm. If the length of the poster is to be 1.02 m, find the breadth of the poster.

   It makes sense to make a sketch and write in the known lengths.

   8 cm

   | Photo | 4.5 cm

   1.02 m

   Poster | Breadth

   First, turn all the units into metres.
   8 cm $= 0.08$ m   4.5 cm $= 0.045$ m
   Now we need to work out the enlargement scale factor.
   Enlargement Scale Factor $= \frac{\text{Large length}}{\text{Small length}} = \frac{1.02}{0.08} = 12.75$
   Now you can work out the breadth of the poster.
   Breadth $=$ Enlargement Scale Factor $\times$ Small breadth $= 12.75 \times 0.045 = 0.57$ m (2 d.p.)

Non-Calculator

3. Find the height of the triangle if the base is 6 cm and the area is 60 cm^2.

   The area formula for a triangle is
   Area $= \frac{1}{2} \times$ base $\times$ height, so
   $$60 = \frac{1}{2} \times 6 \times \text{height}$$
   $$60 = 3 \times \text{height}$$
   $$\text{height} = 60 \div 3 = 20 \text{ cm}$$

6 cm

Calculator

5. a) Find the volume of the cuboid.

2.5 cm 8 cm 2 cm

The formula for the volume of a cuboid is V = length × breadth × height
V = 8 cm × 2 cm × 2.5 cm = 40 cm³

b) Find the surface area of the cuboid.

2 × ((2.5 cm × 8 cm) + (8 cm × 2 cm) + (2.5 cm × 2 cm)) = 2 ×(20cm² + 16cm² + 5cm²)
Surface Area   = 2 × 41 cm²
Surface Area   = 82 cm²

Calculator

6. Find the values for a and b

The angle a will be a right angle, as the triangle is drawn within the circle and one of the triangle's lengths is the diameter of the circle. So, $a = 90°$

To find the length of b, trigonometry will have to be used.
We need to decide what ratio to use (SOHCAHTOA).
We know the value of the hypotenuse and we want to know the adjacent so we can use the Cos ratio.
$\text{Cos } x = \frac{\text{Adjacent}}{\text{Hypotenuse}}$
$\text{Cos } 32° = \frac{b}{14} \rightarrow b = 14 \times \text{Cos } 32° = 11.9 \text{cm (1d.p.)}$

Non-Calculator

7. When planning a Duke of Edinburgh walk, Ivan measured the total distance his group has to walk on the map as 28 cm. If the scale for the map is 1cm:500 m, what distance (in kilometres) do the group have to walk?

First find the distance the group have to walk in metres. Each centimetre on the map represents 500 metres on the ground, so multiply the distance on the map by 500.

28 × 500 = 14000 m

Now turn the distance into kilometres. There are 1000 metres in a kilometre, so divide the distance in metres by 1000 to get the answer.

14000 ÷ 1000 = 14 km

Calculator

8. The circle and the square have the same perimeter. What is the length of the square?

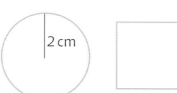

First work out the circumference of the circle.
$r = 2$ cm so $d = 4$ cm
$C = \pi d = \pi \times 4 = 12.6$ cm (1d.p.)
Square's perimeter = 4 × length(cm) = 12.6 cm
Length = 12.6 ÷ 4 = 3.15 cm

Measure and shape

Questions

Calculator

1. The pie chart and table below shows the results from a local election

Labour	28%
SNP	29%
Conservatives	12%
Other Parties	?%

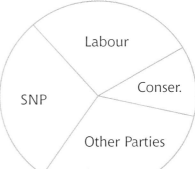

a) What percentage of the vote went to 'Other Parties'?

The total percentage vote is 100%.
Add up the percentages for Labour, SNP and the Conservatives.
28% + 12% + 29% = 69%
Other Parties = 100% − 69% = 31%

b) What will the angle value be in the Labour section?

28% of 360° = 0.28 × 360° = 100.8°

Remember the angles in a circle add up to 360°.

c) 20 000 people voted in the local election. How many voted Conservative?
The Conservatives received 12% of the vote, so:
12% of 20 000 = 0.12 × 20 000 = 2400 people voted Conservative.

Non-Calculator

2. To win a prize in a dart throwing competition, the dart had to land on an odd number.
The numbers that the dart could land on were 2,4,8,12,15,22,46,32,17,9,30.

a) What is the probability of winning a prize?
There are 11 numbers in total and 3 of them are odd. So the probability will be $\frac{3}{11}$.

b) Catriona has 22 throws of the dart. How many times is she likely to win a prize?
$\frac{3}{11} \times 22 = \frac{66}{11} = 6$ prizes

Non-Calculator

3. The numbers below show the results from long jump competition.

2.5m   6.7m   2.4m   5.0m   6.1m   4.4m   5.6m   3.5m   4.0m   4.6m   2.7m
3.1m   2.9m   4.1m   3.9m   6.0m   5.3m

a) Make a stem and leaf diagram of the results.

b) What is the median result?
There are 17 results, the middle result will be the 9th value from the stem and leaf diagram; this is the median.   4.1m is the median.

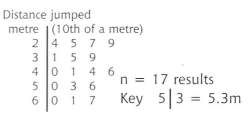

Distance jumped
metre | (10th of a metre)
2 | 4 5 7 9
3 | 1 5 9
4 | 0 1 4 6   n = 17 results
5 | 0 3 6
6 | 0 1 7   Key 5|3 = 5.3m

c) What is the range?
range = highest value − lowest value = 6.7 − 2.4 = 4.3 m = range

Non-Calculator

4.

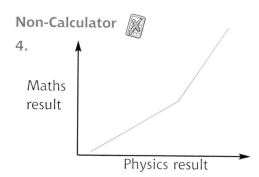

Maths result

Physics result

4. a) What type of correlation does this graph have?
It has a positive correlation, as both variables are increasing.

4. b) Write a sentence that connects the physics results and maths results.
If someone does well in physics it is likely they will do well in maths.

Calculator

5. The Botanical Gardens in Edinburgh have a weather station. The results below show the recorded temperature taken at noon each day during the first week of July.

Mon.	Tue.	Wed.	Thur.	Fri.	Sat.	Sun.
19°C	22°C	20°C	17°C	22°C	23°C	21°C

a) What is the modal temperature?
Remember the mode is the value that occurs the most, so the modal temperature must be 22°C

b) What is the mean temperature for that week?
The mean (or average) is calculated by adding together all the values, then dividing the result by the number of values. There are seven values (one for each day of the week).
$$\text{Mean} = \frac{(19 + 22 + 20 + 17 + 22 + 23 + 21)}{7} = 20.6°C \text{ (1d.p.)}$$

Non-Calculator

6. At the end of the season a football manager works out that his team had won 58% of their matches and had drawn in 18% of them.

a) The team played 50 games that season. How many games did they win?
58% of 50 = 0.58 × 50
= 29 games

b) How many games did the team lose that season?
We need to work out what the percentage is for the team losing. Subtract the percentages for won and drawn matches to get the percentage of matches lost.
100% − 58% − 18% = 24%
Then calculate 24% of the total number of games to find the number of games lost.
24% of 50 = 0.24 × 50 = 12

Calculator

7. Find the mean, median and mode of the following quantities.
12, 17, 22, 18, 19, 26, 14, 19

First put the numbers in numerical order 12, 14, 17, 18,19,19, 22, 24
$$\text{Mean} = \frac{(12 + 14 + 17 + 18 + 19 + 19 + 22 + 24)}{8} = 18.1 (1.d.p.)$$

As there is an even number of values in the set the median will be the average of the middle two numbers. (If there was an odd number of values, the median would simply be the middle number of the set.))
$$\frac{(18 + 19)}{2} = 18.5$$

The mode is 19 as it is the only value to appear more than once.

Mixed questions

1. Audrey paid £3.12 for 13 pencils. Each pencil cost the same. Work out the cost of each pencil.

....

2. Write these numbers in order of size, start with the smallest number: 0.6, 65%, $\frac{1}{2}$, $\frac{6}{7}$, $\frac{3}{8}$

....

3. Rashid carried out a survey to find out the favourite subjects of 24 students. Draw a pie chart of this data.

4. Draw a plane of symmetry on this shape:

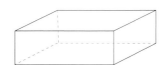

Subject	Frequency
Maths	9
English	4
Art	6
German	5

5. Jonathan is buying a new television. He sees three different advertisements for the same television set. Work out the cost of the televisions in each advertisement.

a)
Ed's Electrical Goods
TVs Normal Price
£250
Sale: 10% off

b)
Sheila's Bargains
TVs £185
+ VAT at 17½%

c)
GITA'S TV SHOP
Normal Price
£290
SALE: ⅓ Off
Normal Price

6. A box of crisps contains 50 packets of different flavoured crisps. There are four flavours: cheese, bacon, beef and tomato, in each box. The probability of each flavour in each box is:

Flavour	Cheese	Bacon	Beef	Tomato
Probability	0.3	0.1	x	0.45

a) Calculate the value of x.
b) Write down the most common flavour of crisp. ....
c) If a packet of crisps is taken out of the box at random
what is the probability that it is either cheese or bacon flavoured? ....
d) For a party Mary buys six boxes of crisps. Estimate how many
packets of crisps will be tomato flavoured. ....

7. Solve the following equations:
a) $2x - 4 = 10$ ....
b) $6x - 3 = 4x + 9$ ....
c) $5(2x + 1) = 20$ ....

8.a) Copy and complete the table below using the equation $y = -2x + 1$

x	2	0	-2
y	-3		

b) On graph paper plot the values for x and y. Join the points with a straight line.
c) Write down the coordinates of the point where the line crosses the x-axis.

 9. A circle has a radius of 36 cm. Work out the circumference of the circle. Give your answer correct to the nearest cm.

....

 10. $s = ut + \frac{1}{2}at^2$. Calculate the value of s when $u = -6$, $t = 4.2$ and $a = \frac{5}{8}$.

....

Indicates that a calculator may be used

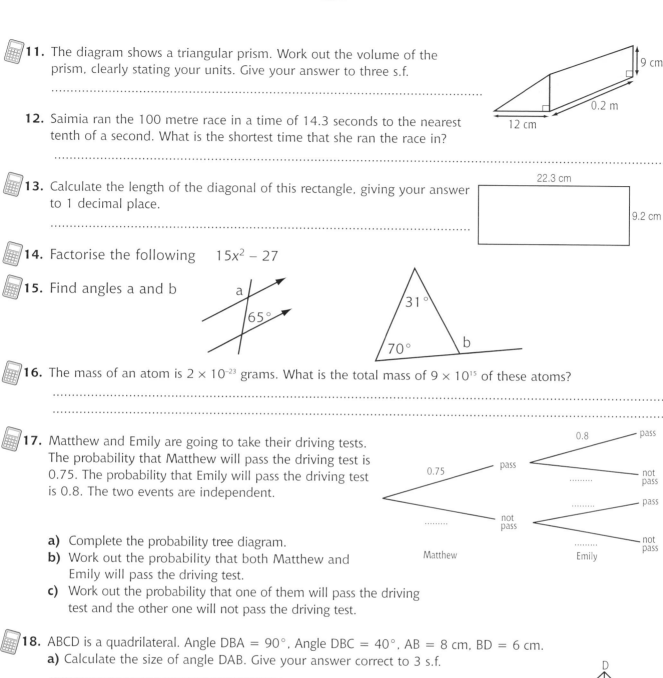

11. The diagram shows a triangular prism. Work out the volume of the prism, clearly stating your units. Give your answer to three s.f.

....

12. Saimia ran the 100 metre race in a time of 14.3 seconds to the nearest tenth of a second. What is the shortest time that she ran the race in?

....

13. Calculate the length of the diagonal of this rectangle, giving your answer to 1 decimal place.

....

14. Factorise the following $15x^2 - 27$

15. Find angles a and b

16. The mass of an atom is 2×10^{-23} grams. What is the total mass of 9×10^{15} of these atoms?

....

....

17. Matthew and Emily are going to take their driving tests. The probability that Matthew will pass the driving test is 0.75. The probability that Emily will pass the driving test is 0.8. The two events are independent.

a) Complete the probability tree diagram.
b) Work out the probability that both Matthew and Emily will pass the driving test.
c) Work out the probability that one of them will pass the driving test and the other one will not pass the driving test.

18. ABCD is a quadrilateral. Angle DBA = 90°, Angle DBC = 40°, AB = 8 cm, BD = 6 cm.
a) Calculate the size of angle DAB. Give your answer correct to 3 s.f.

....

....

b) Calculate the length of DC. Give your answer correct to 3 s.f.

....

....

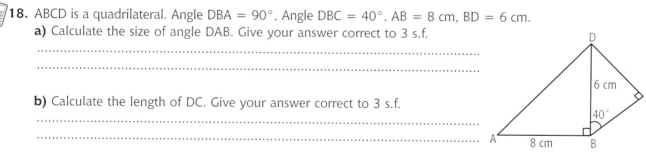

How did you do?

1–5	correct	....start again
6–10	correct	....getting there
11–15	correct	....good work
16–18	correct	....excellent

Answers

Number

1. **a)** 9, 21, 41 **b)** 9, 64, 100
   c) 2, 41 **d)** 2, 40
   e) 40, 64, 100
2. $9°$
3. **a)** 44 764 **b)** 27
4. 8 boxes
5. 77.3%
6. £315

7. 750 g self-raising flour, 375 g butter 625 g sugar, 5 eggs
8. £25 000
9. **a)** 365 **b)** 0.706
10. 104 km
11. The 100 ml tube of toothpaste.
12. **a)** $y = 7.5x$ **b)** $m = 8n^2$
    c) $h = \dfrac{24.5}{j}$

13. £12 500
14. £170.50
15. £5.27
16. 20%
17. **a)** 2.67×10^6 **b)** 4.27×10^3
    c) 3.296×10^{-2} **d)** 2.7×10^{-2}
18. **a)** 4.2×10^{-5} g **b)** 2.65 g
19. $2130.00

Algebra

1. $T = 85y + 8z$
2. **a)** 29, 36 **b)** 106 **c)** $e = 7p + 1$
3. **a)** $x = 3$ **b)** $x = 4$ **c)** $x = 5$ **d)** $x = 2$ **e)** $x = 1$
4. **a)** $9x + 4 = 22$ **b)** 3 cm is the shortest side
5. **a)**

x	−2	−1	0	1	2	3
$y = 3x - 4$	−10	−7	−4	−1	2	5

   b)
   c) $(\frac{4}{3}, 0)$

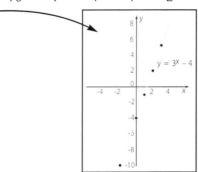

6. $N = 4n + 5$
7. **a)** $2t - 1$ **b)** $18 - 3x$ **c)** $12 + 4w$
8. **a)** $8(2g - 5)$ **b)** $4(3 - x)$ **c)** $j(j + 4)$ **d)** $7q(4q + 3)$
9. **a)** $v = -1$ **b)** $v = -14$
10. $x = 9$
11. **a)** $4 \geq x$ **b)** $14 \geq x$
12. **a)** $m = 4$ $c = 7$ **b)** $m = -\frac{1}{4}$ $c = 3$
    c) $m = -2$ $c = 5$
13. Graph 1 = C Graph 2 = A Graph 3 = B
14. **a)** 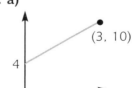 **b)**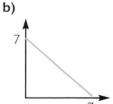

Measure and Shape

1. **a)** ▭ **b)** Order 2
2. $a = 80°$ $b = 50°$ $c = 130°$ $d = 15°$
3. **a)** 74 **b)** 10.75
4. **a)** 3650 m **b)** 3.07 m **c)** 2740 ml
5. 7.5 km
6. $59.5 \leq W \leq 60.5$
7. **a)** 122.5 m² **b)** 63.6 cm²
8. 4 hr 55 min
9. 26.5̇

10. 239.5 miles
11. 5.2 m
12. 1 940 000 cm³
13. 9.2 m
14. 325°
15. **a)** (i) XMN = 83°
    (ii) XNM = XZY = 68° Angles in a triangle add up to 180°, therefore 180° − 68° − 29° = 83°
    b) 3.65 cm
    c) 11.74 cm
16. l $\sqrt{r^2 + t^2}$; πr^2; 2tl

Graphs and Statistics

1. a)

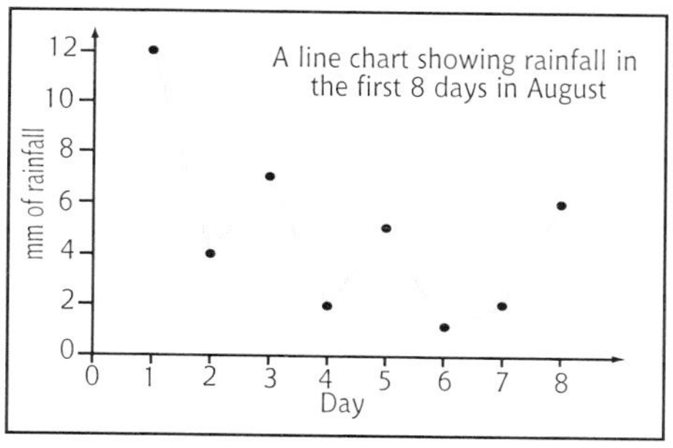

A line chart showing rainfall in the first 8 days in August

(scatter plot: y-axis "mm of rainfall" 0–12, x-axis "Day" 0–8)

b) 4.875 mm

2.

Crisp flavour	No of students	Angle
Cheese & Onion	7	105°
Salt & Vinegar	10	150°
Beef	6	90°
Smokey Bacon	1	15°

(pie chart: Salt & Vinegar 150°, Beef, Cheese & Onion 105°, Smokey Bacon 15°)

3. mean = 3.125
median = 2.5
mode = 2

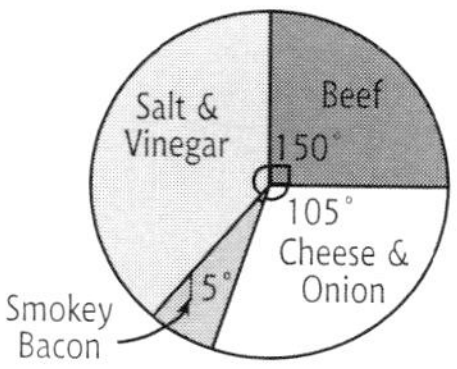

4. a) $\frac{3}{13}$ **b)** $\frac{6}{13}$ **c)** 0 **d)** $\frac{7}{13}$

5. 0.18

6. a)
b) 12 girls

	Under 13 yrs	13 yrs +	Totals
Boys	15	27	42
Girls	12	21	33
Total	27	48	75

7. a) 120°
b) 8 h
c) 6 h

8. 140 people

9. 42 races

10. a) 52 56 60 62 84 85 89 89 91 91 92 95 108
b) 13 hours
c) 81
d) 89

11. a)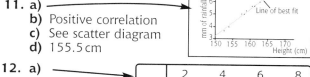
b) Positive correlation
c) See scatter diagram
d) 155.5 cm

12. a)
b) $\frac{16}{16} = 1$
c) $\frac{2}{16} = \frac{1}{8}$

	2	4	6	8
3	6	12	18	24
5	10	20	30	40
5	10	20	30	40
7	14	28	42	56

13. a)
b) 25 kg

5	2	3	5	9
6	0	2	5	7
7	0	7		

14. a)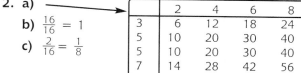
b) 0.51
c) 0.43

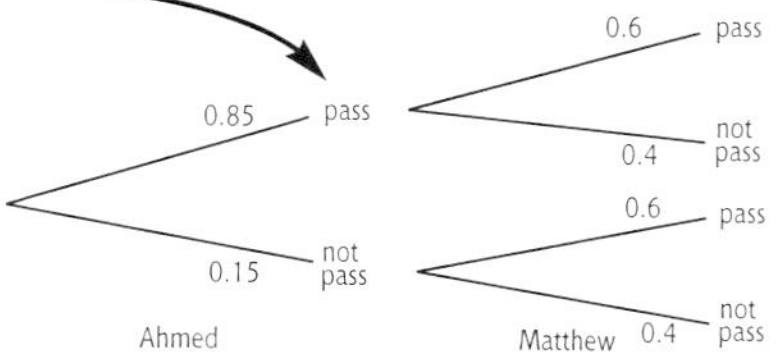

(tree diagram: Ahmed pass 0.85 → Matthew pass 0.6 / not pass 0.4; Ahmed not pass 0.15 → Matthew pass 0.6 / not pass 0.4)

Non-Calculator Skills

1. 210
2. 14.35
3. 484
4. 612
5. $\frac{3}{10}$

6. 24
7. 0.0027
8. a) 4:7 **b)** £4200
9. 123

10. 23% $\frac{1}{3}$ 35% $\frac{2}{5}$ 41% 0.46
11. 17
12. 82°
13. £22

14. 19°C
15. 4.6×10^3
16. 33p
17. 136.8
18. $\frac{1}{15}$

Mixed questions

1. 24p
2. $\frac{3}{8}$, $\frac{1}{2}$, 0.6, 65%, $\frac{6}{7}$
3. 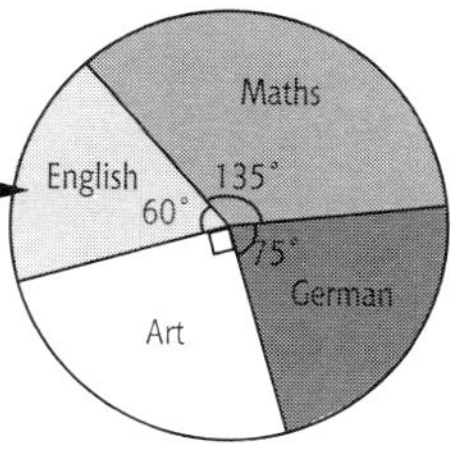
4.

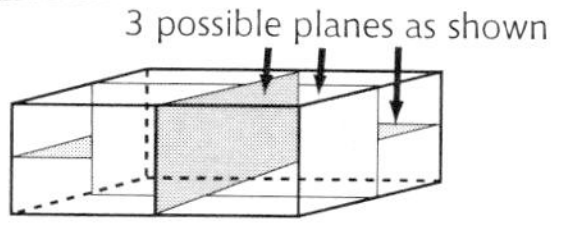

3 possible planes as shown

(pie chart: Maths, English 135°, 60°, 75° German, Art)

5. a) £225 **b)** £217.38 **c)** £232

6. a) 0.15 **b)** Tomato **c)** 0.4 **d)** approx. 135

7. a) $x = 7$ **b)** $x = 6$ **c)** $x = \frac{3}{2}$ or 1.5

8. a)

x	2	0	−2
y	−3	1	5

b) 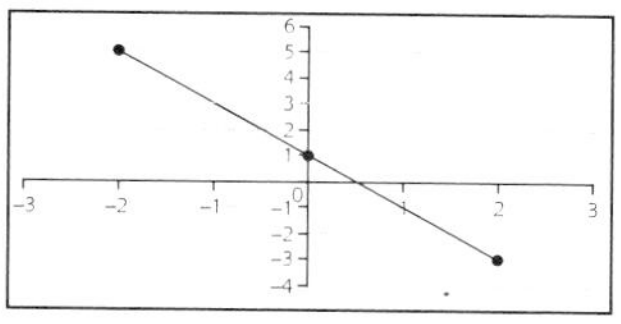
c) $(\frac{1}{2}, 0)$

9. 226 cm
10. −19.6875
11. 1080 cm³
12. 14.25 sec
13. 24.1 cm
14. $3(5x^2 − 9)$
15. a = 115° b = 101°
16. 1.8×10^{-7} grams
17. a) 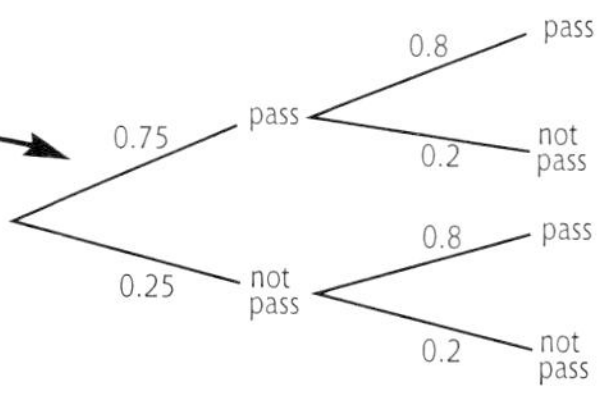
b) 0.6
c) 0.35

(tree diagram: pass 0.75 → pass 0.8 / not pass 0.2; not pass 0.25 → pass 0.8 / not pass 0.2)

18. a) 36.9° **b)** 3.86 cm

Standard Grade Maths (General)

Index

algebra 32–45
angles 47, 53–5, 62–3
   of elevation and depression 64
   see also right angled triangles
approximations 18–19
arcs 52
areas 47, 52
averages 72–3

bearings 56-7
brackets 34,37

calculators, using 7, 8, 17, 29, 33, 62
chords 52
circles 52–3
circumferences 52
class intervals 69
cones 48
continuous data and measures 44, 68–9
conversion graphs 43
correlation 70
cos 62
cube numbers 4, 35
cube roots 4
cubes 46
cuboids 46, 70
cylinders 48

decagons 55
decimal numbers 10–11, 16
   converting to fractions and
    percentages 16
   ordering 10, 16
   rounding 18–19
diameters 52
directed numbers 6
discrete data and measures 44, 72
distance-time graphs 26–7

enlargements 58
equations 36–7
equilateral triangles 47
estimates 18
exhaustive events 74
expected numbers 75
expressions 33

factorisation 34
factors 4, 34
Fibonacci sequences 35
formulae 33
fractions 8–9
   converting to decimals and
    percentages 16
   equivalent fractions 8
frequency 69
frequency tables 69, 73

gradients 40–1
graphs 26–7, 38, 40–3
   curved graphs 43
   scatter graphs 70-1
   straight line graphs 40–3
grouped data 68–9

heptagons 55
hexagons 55
highest common factors (HCFs) 5
histograms 69
hypotenuses 60–3

independent events 76
indices *see* powers
inequalities 38
integers 6
intercepts 41
interest 14
isosceles triangles 47

kites 47

like terms 32
limits 44
line graphs 68
linear equations 36–7
lines 40–3, 54, 61, 71
lines of best fit 71
lines of symmetry 51
lowest common multiples (LCMs) 5

maps 57
means 72–3
medians 72–3
metric units 46
modes 72–3
multiples 5
mutually exclusive events 74, 76

nets 50
nonagons 55
numbers 4–7, 35
   negative numbers 6–7, 34, 38
   ordering 10, 16
   patterns 35

octagons 55

parallel lines 41, 54
parallelograms 47
pentagons 55
percentages 12–16
   converting to decimals and fractions 16
   increase and decrease 12–3
perimeters 37
pie charts 68–9
plane symmetry 51
polygons 55

powers 17, 28–9
   standard form 28–9
   see also cube numbers, square
    numbers
prime factors 5
prime numbers 5
prisms 48
probability 74–7
pyramids 48
Pythagoras' theorem 60–1
quadrilaterals 55

radius 52
ranges 72–3
ratios 20–1, 62
reciprocals 5
rectangles 47
recurring decimals 11
reflective symmetry 51
relative frequencies 75
rhombuses 47
right-angled triangles 47, 60–4
rotations 51

sample space diagrams 77
scale drawing 56–7
scale factor 58
scalene triangles 47
scatter diagrams 70–1
sector 52
segment 52
sequences 35
significant figures 18
similarity 58
sin 62
spheres 48
square numbers 4, 35
square roots 4
squares 47
stem and leaf diagrams 68
substitution 33
symmetry 47, 51

tan 62
tangents 52
terms 32, 35
tessellations 55
tetrahedrons 48
trapeziums 47
travel graphs 26–7
tree diagrams 76–7
triangles 47, 54, 58, 60–4
   length of sides 60, 63
   see also right-angled triangles
triangular numbers 35
trigonometry 62–4

volume 48